CIVIL

MAY 2008
VOLUME 161
SPECIAL ISSUE ONE

ENGINEERING

CONTACTS

Editor:
Simon Fullalove
tel: +44 20 7665 2448
fax +44 20 7538 4101
email: editor@ice.org.uk

Editorial coordinator:
Ben Ramster
tel: +44 20 7665 2242
email: ben.ramster@ice.org.uk

Publisher:
Leon Heward-Mills
tel: +44 20 7665 2450
email: leon.heward-mills@thomastelford.com

Advertising:
Warners Group Publication
The Maltings, West Street, Bourne,
Lincolnshire PE10 9PH
tel: +44 1778 395 029
fax +44 1778 392 079
email: michaelt@warnersgroup.co.uk

Reprints:
Terri Harding
tel: +44 20 7665 2447
email: terri.harding@thomastelford.com

Published by
Thomas Telford Ltd
1 Heron Quay, London E14 4JD
tel: +44 20 7665 2447
fax +44 20 7538 4101
email: journals@thomastelford.com
www.thomastelford.com

Thomas Telford Ltd is a wholly
owned subsidiary of the
Institution of Civil Engineers

Production editing by Richard Sands
Illustrations by Anthea and Ben Carter
Origination by Kneath Associates, Swansea
Printed in the UK by Woodford Litho, Witham

ISSN 0965-089X (Print) 1751-7672 (Online)

© The authors and the Institution
of Civil Engineers, 2008

Available on-line at
www.civilengineering-ice.com

Subscription Information

Non-members:
Subscription enquiries and notification of
change of address should be sent to the
Customer Services Department, Thomas Telford Ltd,
1 Heron Quay, London, E14 4JD, UK
tel: +44 20 7665 2460
fax +44 20 7537 2529
email: journals@thomastelford.com

Civil Engineering, 4 issues per year
(plus two special issues) 2008 subscription price:
UK £123; Elsewhere £142

Full ICE Proceedings Package,
68 issues per year, 2008 subscription price:
UK £2426; Elsewhere £2777

Members:
Subscription enquiries and notification of changes of
address should be sent to
Membership Registry, Institution of Civil Engineers,
PO Box 4479, London SW1P 3XB, UK
tel: +44 20 7665 2227
fax +44 20 7222 3514
email: subs@ice.org.uk

The papers and articles express the opinions of the
authors, and do not necessarily reflect the views of
the ICE, TTL, or the Editorial Panel.

Photos © BAA Ltd: www.baa.com/photolibrary

CONTENTS

Heathrow's new T5 finally opened in March 2008, 23 years after it was first encouraged by the government and after a record 525-day planning inquiry (page 4)

View of the completed twin rivers diversion and realigned western perimeter road looking north towards the new welcome roundabout (page 28)

Bombardier Transportation Innovia driverless trains in the tracked transit system are a development of those in use at Gatwick and Stansted airports: they are initially running every 90 s over a 20 h period (page 58)

Civil Engineering relies entirely on material contributed by civil engineers and related professionals. Illustrated articles up to 750 words and papers of 2000 to 3500 words are welcome on any relevant civil engineering topic that meets the journal's aims of providing a source of reference material, promoting best practice and broadening civil engineers' knowledge. Please contact the editor for further information.

ice Institution of Civil Engineers

PROCEEDINGS OF THE INSTITUTION OF CIVIL ENGINEERS

IN THIS ISSUE

Heathrow Terminal 5: gaining permission

When Terminal 5 at London's Heathrow airport opened on 27 March 2008, it was almost 23 years since the publication of the 1985 Airports Policy White Paper that had encouraged its development. **Roger Pellman** explains the challenges inherent in trying to deliver a major infrastructure project through the UK planning system, and describes how the absence of clear and updated government policy contributed to the record 525 days it took for planning consent to be granted for the £4·3 billion airport expansion. It also recognises that the project involved issues of national importance that affected many thousands of people, particularly those living under the flight paths (pages 4–9).

Heathrow Terminal 5: delivery strategy

The success of the five-year construction phase of Terminal 5 was dependent on putting into effect the principles of a unique form of contract, called the T5 agreement. **Andrew Wolstenholme**, **Ian Fugeman** and **Fiona Hammond** report on how the £4·3 billion scheme required the client to lead in areas that were typically the traditional domain of suppliers or contractor organisations, resulting in novel methods and relationships. From the early stages it was recognised that the project had to be delivered differently to the norm if it was to achieve its desired objectives. This resulted in the phrase that all those involved in the delivery of the T5 were part of 'history in the making' (pages 10–15).

Heathrow Terminal 5: health and safety leadership

T5 has gained widespread recognition for a safety performance some four times better than industry norms and for setting new benchmarks for occupational health and safety. The sheer size of the £4·3 billion project, its structural and managerial complexities and the 60 000 people that have worked on site have presented significant challenges. As **Mike Evans** describes, safety leadership—as distinct from safety management—coupled with real engagement of, and respect for, all concerned has led to cultural change. This has not only reduced the number and severity of injuries but has also resulted in improved worker health, satisfaction, morale and performance (pages 16–20).

Heathrow Terminal 5: enhancing environmental sustainability

T5 provided an opportunity for owner British Airports Authority to set and deliver new standards in environmental sustainability for the construction industry. **Beverley Lister** shows how opportunities were pursued at each development stage to improve performance and firmly embed environmental awareness and corporate responsibility into decision-making processes. Throughout the design and construction, project teams and suppliers were incentivised to apply innovative techniques and best practice to deliver exemplary environmental performance (pages 21–24).

Heathrow Terminal 5: twin rivers diversion

Moving two rivers from the middle of the T5 site to a new alignment around the western airport perimeter was a critical and highly environmentally sensitive sub-project of the T5 programme. The rivers diversion had to be completed before the original river structures could be demolished to enable continuation of the main terminal development. As described by **David Palmer**, the £45 million diversion project included creation of two 3 km long river channels, phased realignment of 3 km of highway and landscape works to the western boundary of Heathrow (pages 25–29).

Heathrow Terminal 5: tunnelled underground infrastructure

T5 was one of the largest infrastructure projects in Europe over the last five years. It included approximately 14 km of tunnelling to provide underground infrastructure for rail, road and effluent discharge. In all cases the tunnels were delivered without major incident, thanks to a single integrated team which made all aspects of risk management central to its delivery philosophy. **Ian Williams** outlines the four key projects for which the key aspect of delivery was tunnelling – the airside road, the storm water outfall and the Piccadilly line and Heathrow Express extensions. It discusses the range of aspects that made up the delivery of this infrastructure, ranging from design through to performance of tunnelling machines and the control of surface settlement (pages 30–37).

Heathrow Terminal 5: building substructures and pavements

Tim Dawson, **Kathiresapillai Lingham**, **Roger Yenn**, **Jim Beveridge**, **Richard Moore** and **Matthew Prentice** describe the design and construction of the vast piled basement structures for the three terminal buildings at the £4·3 billion T5 project, together with 1 million m^2 of associated aircraft pavements. The basements are up to 20 m deep and involved the excavation and reuse of 6·5 million m^3 of gravel and clay. The aircraft pavements involved a number of innovations including development of a new high-strength concrete, which delivered a thinner construction and resulted in programme and environmental benefits (pages 38–44).

Heathrow Terminal 5: terminals T5A and T5B

Steve McKechnie, **Dervilla Mitchell**, **William Frankland** and **Maurice Drake** describe the design and construction of the two terminal buildings in phase 1 of T5. A total of 40 000 t of steel was used to create 280 000 m^2 of space in the dramatic main terminal building, T5A, including forming a clear span roof of 156 m × 396 m to enclose it. In addition to high visibility for passengers, the design provides maximum flexibility for future modifications. In the first of two satellite buildings, T5B, 600 000 m^2 of post-tensioned flat slab were cast using minimal amounts of formwork and site labour. Building these vast structures at the world's busiest international airport also meant all construction operations had to be undertaken within a highly restricted space and with no cranes allowed above roof level (pages 45–53).

Heathrow Terminal 5: rail transportation systems

Critical to opening T5 was the availability of a public service on extensions to both the Heathrow Express and London Underground's Piccadilly line, not least as this was a key requirement of the planning permission. A sub-surface tracked transit system also needed to be operational to provide a passenger link between the two main terminal buildings. **Ian Fugeman** explains the challenge of managing the design, delivery and integration of new railway systems into existing, occasionally old, public systems with continually evolving management structures and without impacting on existing rail services. This was greatly assisted by the application of a system engineering approach (pages 54–59).

Heathrow Terminal 5: energy centre

Heating and cooling the vast new terminal buildings at T5 development is undertaken by a dedicated energy centre, which provides continuous supplies of hot and chilled water for heating and air-conditioning respectively. The hot water is provided by a combination of a local combined-heat-and-power source and natural gas boilers, which can also run on fuel oil, and the chillers are powered by high-voltage electricity. **George Adams** describes the design and construction of the highly efficient centre, which also made extensive use of off-site testing and manufacturing (pages 60–64).

Introduction

BAA Ltd successfully completed its £4·3 billion Terminal 5 (T5) development at London's Heathrow airport on time and to budget on 27 March 2008. This huge project's planning history dates back nearly a quarter of a century and, during its five-year construction programme, was one of the largest construction schemes in Europe.

The sheer scale and complexity of the T5 programme means selecting ten papers to publish in this special issue of the Institution of Civil Engineers' *Civil Engineering* journal has been no easy task. The 16 projects that made up this vast and varied programme of works and the countless engineering and delivery challenges that have been met simply cannot be done justice in just 64 pages.

Nevertheless, we hope the papers presented here will provide a useful overview of the history of the T5 programme; of the strategic approach taken to managing risk, safety and sustainability; and of the design and construction of the key civil, structural and other engineering elements.

Roger Pellman starts by explaining the challenges inherent in trying to deliver a major infrastructure project through the UK planning system. In recent years the media has rightly challenged the adequacy of Heathrow, its passenger handling difficulties and its ability to cope with the UK government's ever-changing response to the threat of global terrorism. But when T5 opened in March 2008 it was almost 23 years since the publication of the 1985 Airports Policy White Paper which encouraged its development – something which needs resolving if future projects of national importance are to be delivered more expeditiously.

The next three papers address three important aspects of 'how' the programme was delivered. The chosen procurement strategy was in direct response to the assessed risk that the delivery of a programme of such magnitude presented to BAA as a business. Many lessons can be learnt from the innovative approach adopted but the integrated team, the communication of a clear set of values and alignment of all parties are a consistent theme – as seen in the paper by Ian Fugeman and myself on delivery strategy; the paper by Mike Evans on health and safety; and the paper by Beverley Lister on enhancing environmental sustainability.

When the works started on site in late summer of 2002, one of the most critical activities was to relocate the existing Duke of Northumberland's river and Longford river, which ran directly across the site, to a new location along the airport's western perimeter. The 'twin rivers' were a major topic in the long-running planning enquiry and BAA worked closely with the Environment Agency to establish a viable scheme, which – as can be seen from David Palmer's paper – has achieved significant eco-system and habitat gain.

BAA had decided in 2000, well before planning permission had been granted for T5, to proceed with the construction of 1·4 km airside road tunnel to improve access to the existing remote aircraft stands situated at that time at the western perimeter of the airport. The road tunnel also forms part of the airside road system which now connects the central area of Heathrow airport with the new T5 facility. The team established for this project continued with the delivery of all bored tunnelling which by 2006 had totalled nearly 14 km. Ian Williams' paper provides an overview of all the tunnelling activity, which was also a particularly successful example of an integrated team benefiting from the 'T5 agreement' form of contract.

Some of the first structures to be built on the site were the extensive sub-surface rail boxes, which extend to the western boundary of the site, and the large basements for the terminal buildings, which are up to 20 m deep. These and associated civil works – including pavements for the new generation of large aircraft – are described by Tim Dawson, Kathiresapillai Lingham, Roger Yenn, Jim Beveridge, Richard Moore and Matthew Prentice.

Steve McKechnie, Dervilla Mitchell, William Frankland and Maurice Drake then describe the design and construction of the superstructures for the gigantic main T5A terminal building and its satellite T5B, which on its own is larger in plan area than Terminal 4. They provide an insight in their paper into the brilliant response that both the engineers and the architects brought to the challenge of achieving a pleasurable space with operational flexibility but delivered to a tight construction programme and executed within a very constrained site.

The mechanical and electrical systems required to support the safe and efficient operation of the terminal buildings are so significant that they warrant their own special edition, possibly by other engineering institutions. The last two papers in this issue address two of these – the three railway systems and the energy centre.

The extension of BAA's Heathrow Express service to Heathrow together with the extension of London Underground's Piccadilly line were vital elements of the T5 surface-access strategy but were also conditions of the planning permission. The management of the complex approval activity for these two extensions and the interface with the numerous operators and third parties through a variety of contractual arrangements are described in Ian Fugeman's paper. The third railway system is the tracked transit system, a driverless vehicle operating below ground which carries passengers between the main terminal building and the satellite buildings.

George Adams concludes the issue with a description of T5's dedicated and highly efficient energy centre, which provides continuous supplies of hot and chilled water to the termi-

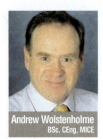

Andrew Wolstenholme
BSc, CEng, MICE

is capital projects director at BAA

nal buildings for heating and air-conditioning respectively.

Many other technical papers have or will soon be published which provide further description of the various technical challenges. These include Richard Matthews' paper on the new 87 m air-traffic control tower – the most visible structure at Heathrow and created as part of the T5 programme – which is published alongside this special issue in the normal May 2008 issue of *Civil Engineering*;[1] and William Frankland's paper on delivery of the T5A roof in the September 2007 issue of ICE's *Management Procurement and Law* journal.[2]

While the T5 delivery phase lasted some five years it was most probably in the preceding years that the foundations for this remarkably successful project were established. The formation of integrated teams and the application of a novel form of contract enabled trust between all parties to be established. This and the importance of effective leadership at all levels will be remembered by all those who were involved in the making of T5.

My thanks go to Ian Fugeman, for all his efforts in coordinating this special issue; to the authors, who gave their own time to record and share their experiences on T5; and to the volunteer ICE referees and assessors for peer-reviewing the papers. I hope you enjoy reading them.

References

1. MATTHEWS R. Creating Heathrow's new eye in the sky. *Proceedings of the Institution of Civil Engineers, Civil Engineering*, 2008, **161**, No. 2, 66–76.
2. FRANKLAND W., KITCHENER J. N., WHITTEN T. and HULME P. The delivery of the roof of Terminal 5. *Proceedings of the Institution of Civil Engineers, Management, Procurement and Law*, 2007, **160**, No. 3, 101–115.

What do you think?

If you would like to comment on this article, please email up to 200 words to the editor at journals@ice.org.uk.

If you would like to write a paper of 2000 to 3500 words about your own experience in this or any related area of civil engineering, the editor will be happy to provide any help or advice you need.

Proceedings of ICE
Civil Engineering 161 May 2008
Pages 4–9 Paper 700041

doi: 10.1680/cien.2007.161.5.4

Keywords
airports; government; infrastructure
planning

Roger Pellman
BA

is BAA Ltd's head of environmental
planning for the Stansted generation
2 (new runway) project and formerly
general manager of the Heathrow
Terminal 5 inquiry for BAA

Heathrow Terminal 5: gaining permission

When Terminal 5 at London's Heathrow airport opened on
27 March 2008, it was almost 23 years since the publication of
the 1985 Airports Policy White Paper that had encouraged its
development. This paper explains the challenges inherent in trying
to deliver a major infrastructure project through the UK planning
system, and describes how the absence of clear and updated
government policy contributed to the record 525 days spent at
the planning enquiry before consent for the £4·3 billion airport
expansion could be granted. It also recognises that the project
involved issues of national importance that affected many thousands
of people, particularly those living under the flight paths.

When Terminal 5 (T5) at London's Heathrow
airport opened on 27 March 2008 (Fig. 1), it
was almost 23 years since the publication of the
Airports Policy White Paper[1] that had encouraged its development. This fact, and the reasons
for it, have resulted in a number of overhauls
and improvements to the way the UK planning
system deals with major infrastructure projects
in order to make the system quicker, simpler

and more inclusive. The Planning White Paper,[2]
published in May 2007, is the latest example of
the government trying to inject some momen-
tum into what had become a moribund process.

This paper identifies some of the reasons why
the T5 project took so long to deliver. Some of
them are good reasons. For example, finding
the right solution to the loss of the Perry Oaks
sewage sludge treatment works on which the

Fig. 1. Heathrow's new T5 finally opened in March 2008, 23 years after it was first encouraged by
the government and after a record 525-day planning inquiry

Institution of Civil Engineers

project is built took time, as did finding the right surface-access infrastructure solutions which the then Department of Transport (DoT) could support. Its passage through the planning system involved issues of national importance, which touched upon the lives of many thousands of people, particularly those living under the flight paths.

Other reasons are not so good. The 1985 White Paper was, at best, equivocal in its support for T5; it was said to be 'beginning to look a little yellow at the edges' by the then minister for aviation Lord Caithness even before the public inquiry started in 1995. The inquiry itself lacked clear statements of national policy on the need for more airport capacity generally and at Heathrow in particular, and on the government's approach to air noise and air quality. The absence of clear and updated policy added many months to the inquiry.

Planning and policy framework

Major airport development proposals are made by airport operators against a background of national, regional and local planning policies. The post-World War II period has been characterised by statements of government policy (mostly in the form of White Papers) supporting a planning-led approach to airport development.

Some White Papers on the future of the air transport industry in this period have been better than others. However, the fact that airport development can give rise to significant local environmental effects, as well as providing substantial regional and national economic benefits, underpins the government's approach that national policy statements should direct where and when airport development takes place. This approach is to be preferred because the impacts of an airport will extend well beyond its boundary, and such consequences need to be covered within the delivery plans of other agencies. For example, local authorities need to plan for population and employment growth, for airport-related land uses and local road provision; the strategic highway and rail authorities need to consider new surface-access infrastructure and plan programmes and budget delivery accordingly.

The 1978 Airports Policy White Paper[3] concluded that, for the handling of air passenger demand in the south east of the UK up to 1990

- the proposed new airport at Maplin Sands in the Thames estuary (east of London) would be abandoned
- BAA proposals for Terminal 4 at Heathrow (west of London) should be examined at a public inquiry
- BAA should bring forward proposals for a second passenger terminal at Gatwick airport (south of London)
- BAA should bring forward proposals at

Stansted airport (north east of London) to handle some 4 million passengers a year
- Luton airport (north of London) should be restricted to 5 million passengers a year.

This strategy was based on the recognition (in paragraph 153 of the White Paper[3]) that the air transport industry

'makes a valuable contribution to the economy; the principle that air transport facilities should not in general be subsidised by the taxpayer or ratepayer; the need to take careful account of the environmental impact of airport developments; the adoption of a flexible approach in planning and development which can be related to growth in demand; and the importance of consulting those concerned with airports in the development of policy'.

These five principles run like a thread through airports policy statements ever since 1978 and feature prominently in the White Papers of 1985[1] and 2003.[4] They were the principles that guided BAA's approach to the planning of T5.

The 1985 Airports Policy White Paper followed the 1981–1983 public inquiries into proposals for the development of Stansted and Heathrow airports. At those inquiries, the BAA planning application to develop Stansted—submitted at the invitation of a national policy statement by the government in December 1979[5]—had been opposed by Uttlesford District Council (the local planning authority for Stansted airport) and British Airways (BA), which had then submitted a counter application for the development of a fifth passenger terminal on the Perry Oaks site at Heathrow. BAA argued that there was no prospect of a replacement site for the Perry Oaks function being found in time to meet air passenger demand in the south east and that Stansted provided the only opportunity to do this, even though it was acknowledged that passenger demand for Heathrow was very strong. The inquiries inspector and the Secretary of State agreed with BAA's assessment of the situation and in June 1985 granted consent for development at Stansted.

While the 1985 White Paper concluded, among other things, that

- planning permission had been given for Stansted to grow to a capacity for about 15 million passengers a year
- in the future, demand may justify Stansted's growth to 25 million passengers a year
- a second runway at Gatwick should not be constructed

it also concluded, however, that

- an air-transport-movement limit of 275 000

at Heathrow would not be pursued
- improvements to surface-access links to Heathrow should be urgently studied
- BAA and Thames Water should jointly study options for removing the sludge works at Perry Oaks with a view to releasing the site for airport development
- the possibility of a fifth terminal at Heathrow would be kept under review.

Whereas the last group of conclusions was in itself significant, helpful in pointing the direction of government thinking, and helpful in motivating the likes of Thames Water and the road and rail authorities in new ways, the language did not represent a strong and unequivocal endorsement of a fifth terminal at Heathrow.

Building on the 1985 White Paper

Before BAA would begin planning of additional airport facilities on the Perry Oaks site, it needed to satisfy itself on two counts—first, the outcome of the joint Thames Water/BAA study on a Perry Oaks replacement and, second, the findings of the Heathrow surface-access studies.

The Thames Water/BAA study,[6] conducted between July 1985 and April 1986, concluded that the prospect of finding an alternative site that could replace the function of Perry Oaks was viable. At that stage, there was neither a favoured site nor a favoured de-watering process, but Thames Water and BAA agreed that sufficient prospects of success existed for the next stage of the study to commence. A Heathrow surface-access working group was set up in August 1985 and reported its findings in June 1987.[7] The DoT's conclusions on the findings were published in May 1988. These conclusions led directly to the heavy-rail Heathrow Express service direct to Heathrow from Paddington and to improvements to the A4, M4 and M25. Importantly for BAA, the conclusions were owned by the then DoT and the solutions had DoT support.

In spring 1988, BAA commissioned engineering consultancy WS Atkins to begin airport-access feasibility studies, initial planning for additional airport facilities commenced and legal counsel was appointed for the T5 project.

Initial planning studies

When planning for T5 started in February 1988, the first task was to establish the bounds of the available site. To the north, east and south, the boundaries were defined by the layout of the airfield taxiways. The boundary to the west was more variable, the principal constraints being the A3044 highway (immediately to the west of the runways and the airport perimeter road), the River Colne (midway between the A3044 and the M25 motorway) and, as a hard constraint, the M25 itself (Fig. 2). At this stage, BAA did not impose upon itself any limitation on the land that might be available to the west of Heathrow, except for the presence of the M25.

BAA had to form a judgement about the maximum size of aircraft to operate at Heathrow in order to define airfield boundaries. At the time, the B747 had a 60 m wingspan, which would increase to nearly 65 m with the introduction of the B747-400. Plans were under consideration by major aircraft manufacturers for a new generation of larger aircraft with significantly larger wingspans. BAA reviewed these plans and decided to extrapolate existing taxiway-separation criteria to define a taxiway box around the T5 site that would protect the operation of an aircraft with a wingspan of up to 85 m (the A380-800 has been developed with a wingspan of 80 m).

Early studies in 1988 identified the idealised basic form of T5 as a core terminal building with wrap-around contact stands, augmented by two or possibly three linear satellites accessed by underground links for passengers, luggage transport and vehicles.

Planning of the Perry Oaks replacement facility and the development of surface-access infrastructure was proceeding independently. Thames Water began its own study, funded by BAA, to identify an alternative site in February 1989 and, in the same year, BAA began funding the clearance of sludge from the Perry Oaks site to local farmland. Thames Water's preferences changed over time as government policy, water-treatment processes and the attitudes of landowners and food retailers shifted. Thames Water settled on its preferred dewatering process at Iver South in June 1991 and completed its planning applications for the site and related pipework in November 1994. By that time, BAA had paid for Thames Water to install centrifuges on the Perry Oaks site in order to accelerate the rate at which the Perry Oaks product could be removed, with the aim of clearing the site by 1998.

Transport and Works Act orders for the extension of Heathrow Express were published in September 1994 and for the extension of the Piccadilly underground line in November 1994. Roads orders for the T5/M25 spur road were published in May 1996 and a scheme for improvements to the M4 was published in March 1997.

By spring 1988, having been satisfied of the prospects of acquiring the land and delivering appropriate surface-access solutions that would be needed if consent was to be granted, BAA began to consider the business case for development, what form the development should take and how it should go about planning the facilities and undertaking an environmental assessment. The key questions were around

■ the south-east England and Heathrow passenger and aircraft demand forecasts
■ assumptions about development at other airports in the south-east
■ the period over which BAA should plan its investment
■ the future capacity shortages that any development would need to address
■ whether there were other objectives this additional capacity should seek to deliver.

As a result, the important principles that were established by the beginning of 1990 were

■ no new runway capacity was likely to be permitted in the south-east until best use was made of the existing runways
■ the passenger-handling capacity of the two runways at Heathrow, operated in segregated mode, was far greater than the passenger-handling capacity of the existing passenger terminals and aircraft stands
■ airport and airline advantages could be realised with a strategy for occupation of T5 by an airline or airlines that maximised the potential for redeveloping and refurbishing Heathrow's central terminal area (comprising Terminals 1, 2 and 3) as T5's facilities became available.

Other fundamental issues

Heathrow is the UK's major international airport and, as demand rises, so the airport is increasingly vulnerable to foreign competition as major European airports expand at Heathrow's—and the UK's—expense. Heathrow supports billions of pounds of British exports, thousands of UK jobs and encourages hundreds of international businesses to locate in the UK. Providing more capacity was fundamentally and unequivocally in the national interest.

Planning application for T5

Once BAA had taken the decision to provide principally more passenger terminals and more aircraft-handling capacity, the main decision in the development was whether to extend the airport west to the M25 or to contain the development between the two runways and within the existing airport perimeter road. A review of the options in 1991 concluded, among other considerations (operations, cost, environment and legal), that two key features of the concept, namely

■ retention of the ability to opt for a single-level terminal concept (as at Stansted) with a larger floorplate in the east–west direction rather than a multi-level terminal concept (as at Terminal 4)
■ the location of long-term car parking at surface level adjacent to the terminal building

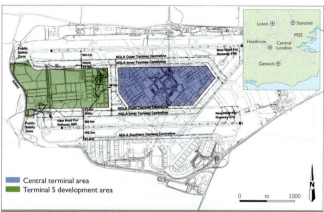

Fig. 2. Heathrow location (inset) and site plan just before the inquiry started in April 1995, showing the 126 ha T5 development area on the former Perry Oaks sewage sludge treatment works and the development limits imposed by taxiways for new-generation large aircraft (NGLA) with wingspans up to 85 m

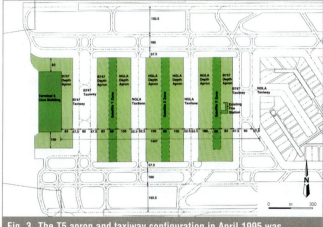

Fig. 3. The T5 apron and taxiway configuration in April 1995 was designed for maximum flexibility, including future use of satellite terminals by new-generation large aircraft (NGLA) (dimensions in m)

provided insufficient justification for development in the Colne Valley regional park and green belt to the west of Heathrow.

BAA's objective therefore became to maximise the additional capacity that could be provided from the acquisition of the Perry Oaks site. Maximising the capacity of the site became of paramount importance given that demand was forecast greatly to exceed capacity at Heathrow, even with T5, and was a key argument for BAA refusing to retain a smaller sludge works site on the redeveloped airport. The most significant effects of this decision were to

- reduce the number of cross-taxiways on the site
- require the adoption of a multi-level terminal concept
- decant much of the T5 car parking to an alternative site within Heathrow's boundary on the north side of the airport.

In March 1992, the BAA board approved, in principle, the scheme that became the basis of the later planning application (i.e. maximising the site produced a plan that would deliver additional handling capacity for about 30 million passengers a year) and further, more detailed, environmental assessment studies began.

BAA announced its intention to apply for consent for T5 in April 1992 and, between May 1992 and January 1993, it made public presentations about its plans to local, regional and national audiences. Consultation documents on the proposals were published in June 1992.

Leading counsel presented his formal opinion on the proposals to BAA in December 1992. The T5 application was submitted in February 1993, and 'called-in' a month later by the secretary of state for transport for his determination. An outline statement of case was submitted by BAA in September 1993.

At this stage, the details of the master plan for T5 continued to be developed as planning and design for T5 progressed. Building flexibility into the master plan to enable it to cope with change was an important consideration, particularly given T5's long gestation period. This is apparent in the conceptual configuration of the rectilinear airfield layout (Fig. 3) in which

the long north–south runs of the aircraft parking apron are capable of supporting a wide range of different aircraft stand arrangements. It is also apparent in the cross-section of the multi-level terminal building. This was planned as a simple, large box with aircraft stands on three faces. Within it, the vertically stacked layout of departing passenger flows above arriving passenger flows has evolved considerably over time within the overall envelope—the configuration by March 1996 is shown in Fig. 4.

Access to and from T5 has been more tightly defined due to the links to third-party road and rail infrastructure. The terminal is accessed by means of a new motorway spur to the M25 and by links to the airport perimeter road. The Heathrow Express and Piccadilly line have both been extended from the central terminal area to T5 and provision is made for further rail access from the west, most probably in the form of the Airtrack scheme. A key feature of the master plan was the creation of a high-quality public transport interchange to encourage and facilitate travel to, from and via Heathrow by all modes of public transport.

The planning inquiry

An inquiry inspector was not appointed by the secretary of state until 29 March 1994, over a year after the application was submitted. He then held five pre-inquiry meetings before the inquiry opened in May 1995. It was then expected to last about 18 months.

In the two years since the planning application—and in addition to the outline statement of case, the road and rail orders and the Thames Water applications already mentioned—BAA submitted a further nine planning applications and a statement of case. By the time the inquiry closed, the number of planning applications and draft orders on which the inspector had to report had increased to 37.

The inquiry took 525 days, between 16 May 1995 and 17 March 1999, making it the longest in UK planning history. The previous record of 340 days was held by the Sizewell B nuclear power station inquiry. During this period, the T5 inspector heard 734 witnesses and was presented with 5900 inquiry documents. Over 20 million words were spoken and recorded on a

daily transcript. The inspector also carried out around 100 site inspections and sat at 18 public meetings at locations around Heathrow, at which over 300 members of the public gave evidence.

The inspector was assisted throughout the inquiry by a deputy inspector; eight other inspectors also sat with him for particular topics. He was supported by a secretariat, administrative assistants and transcribers of the evidence.

The inquiry was held on a topic basis, by agreement with all the main parties. It began with the need for T5, followed by development pressures and socio-economic impacts and then development plan policies and land-use considerations. These general topics were then followed by specific topics such as surface access, noise, air quality, public safety and construction. The need case occupied 123 days of inquiry time, surface access took 117 days and noise lasted 73 days.

Concerns of the local community

The provision of additional capacity at Heathrow was forcefully opposed by local opposition groups, in principle on grounds of policy, need and sustainability. Behind that position, the principal concerns of the community were air and ground noise, air quality, surface access, visual intrusion, risks to public safety, loss and damage to the green belt and to the Colne Valley regional park, and ecological harm.

By far the greatest concern related to air noise. The inspector noted that aircraft using Heathrow caused disturbance and annoyance over a wide area, and that the very great increase in the number of aircraft using Heathrow had made the noise climate worse, particularly in the early morning. Night noise was also regarded as an unnecessary burden on the local communities. It was also asserted that the method of producing noise contours (which describe exposure to different levels of noise) significantly underestimated the contribution that numbers of aircraft made to the overall impact. A limit on the number of aircraft movements and a noise cap were inevitable and they have since become routine features of consents for major airport expansion.

There was evidence that T5 would increase concentrations of pollutants around Heathrow.

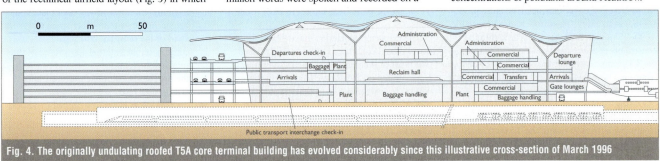

Fig. 4. The originally undulating roofed T5A core terminal building has evolved considerably since this illustrative cross-section of March 1996

Yet, because emissions from road traffic generally and individual aircraft were expected to fall, pollution was predicted to be less in 2016 than it was in 1993, even with T5 operating. T5 would lead to a small worsening of that overall improvement.

Surface access to Heathrow and Paddington was an issue for many. The road and rail infrastructure strategy proposed by BAA was thought to be capable of dealing with arising problems, although the inspector did reinforce the strategy through the imposition of a cap on car parking numbers.

T5 was regarded as visually intrusive. These concerns were mitigated by the quality of the design of the principal buildings. However, the inspector responded to objections by reducing the scale of the multi-storey car parks in such a way as to emphasise the passenger terminal and permit more landscaping.

In terms of public safety, it was considered that there was not an increase in those exposed to intolerable risk.

Less was said about loss or damage to the green belt (within which the Perry Oaks site was located). The impact on the Colne Valley regional park was judged in the context of its landscape character, which was regarded as largely undistinguished.

Ecological harm was not substantial, but did result in costly mitigation. The most significant was a decision taken by BAA during the inquiry to address objections from the then National Rivers Authority (later the Environment Agency) by changing the diversion design of two local rivers. The rivers ran in inverted siphons beneath each main runway and in open channel between them through the Perry Oaks site. It had been planned to link the siphons by placing the rivers in culvert beneath the T5 development. In the face of intense objection by the Environment Agency, BAA changed the proposed solution to a diversion of both rivers in open channel around the ends of the runway and through the already intensively used landside terminal zone.

Airport user issues

While the inquiry was in progress, BAA was also trying to resolve which airline or airlines should occupy T5, and how it was going to work with any potential occupant to design world-class passenger facilities. The basic proposition was to enable Heathrow to grow its traffic to 90–95 million passengers a year within the constraints of a two-runway airport. The task of the T5 planning team in the early 1990s was to determine how this additional capacity could be most efficiently and effectively used.

During 1994–1995, an extensive study and consultation exercise was undertaken with the Heathrow airlines to determine T5 occupancy. Four key selection criteria were used to establish suitable options

- maximising use of Heathrow's passenger-handling capacity
- maximising single-terminal transfer connections
- minimising the required number of airline moves
- meeting airline requirements.

BA was identified as the prime airline occupant of T5, with parts of Terminals 1 and 4 would be released for use by other airlines. BA represents around 40% of the passenger traffic at Heathrow and its mix of domestic, short- and long-haul traffic generally offers a flat daily profile of traffic that allows the infrastructure to be used effectively. BA also has high loads of connection traffic, so its location in a single terminal provides a simpler journey for transferring passengers, reduces the risk of missed connections and avoids lengthy journeys across Heathrow to other terminals.

The client team, with input from BA, produced around 300 briefs setting out stakeholder requirements. The team had to manage many challenges, including how to specify, in the 1990s, requirements for a facility that would not be operational until the next decade. Maintaining flexibility while achieving an efficient design was a major consideration. The brief set out a number of key strategies

- public transport interchange
- good way-finding
- flexibility
- sustainability
- accessibility
- business continuity
- supporting punctuality
- efficiency.

These strategies were tested at various stages of the project in order to maintain confidence in the functionality of the facilities. This was particularly important as business requirements changed over time.

The consent process

The inspector delivered his report to the secretary of state on 21 November 2000. The main report[8] ran to 618 pages. It recommended that T5 should be permitted, subject to strict conditions.

The inspector concluded that demand for air travel would continue to grow in the UK, and that most of the demand in the south-east was concentrated on Heathrow. He thought there was no realistic prospect of that demand being met at other airports in the UK, and that runway capacity was a fundamental limit at both Gatwick and Stansted. He believed T5 would be consistent with the government's objective of fostering a strong and competitive British airline

industry, and that the beneficiaries if T5 was not provided would be Charles de Gaulle, Frankfurt and Schiphol. He noted that Heathrow contributed some 1% to the national economy and played a significant part in attracting investment to the UK. Unless Heathrow was able to maintain its competitive position, there was a substantial risk that London's success as a world city and financial centre would be threatened.

He recommended a limit on the number of air transport movements and a noise cap. He also said that BAA should be required to extend the Heathrow Express and the Piccadilly line to T5 and that car parking should be capped. The inspector said that no further major development should take place at Heathrow after T5 and that a third runway would be totally unacceptable.

The main report was confined to matters that the inspector viewed as relevant to the balance he had to strike; fuller statements of the cases presented to him appeared in supporting topic reports, which ran to many thousands of pages.

Planning consent

While the secretary of state was considering the inspector's report, BAA discovered that the scheme for the diversion of two rivers that ran through the site would not maintain existing flows to their downstream destinations. The secretary of state was informed of this in May 2001 and, following urgent discussions between BAA, the Environment Agency, the Royal Parks Agency and local planning authorities, a new rivers diversion scheme was submitted in August 2001.[9]

Consent for T5 was granted, with almost 700 planning conditions, on 20 November 2001.[10] Consent for the other 36 applications was also granted. No permitted development could start until planning permission had been granted by the local planning authority for the permanent diversion of the two rivers. Otherwise, the principal conditions were

- extension of the Heathrow Express and the Piccadilly line to the site
- a limit of 480 000 air transport movements during any period of one year
- a noise contour cap of 145 km^2 (based on the 57dB(A) 16 h contour)
- controls on the siting (by co-ordinates), maximum dimensions (by floor plate and height) and maximum floor space (by gross area) of the principal buildings
- controls on the external appearance of the principal buildings
- the provision of a public transport interchange at T5
- car parking at Heathrow capped to 42 000 spaces.

Following consent, a local kennels operator launched a legal challenge that was dismissed in the High Court in May 2002. In July, the land at Perry Oaks was finally transferred into BAA's ownership. In the same month, BAA was granted consent by the London Borough of Hillingdon for the revised scheme for the diversion of the two rivers, and this allowed preliminary site works to start in August. Construction of T5 finally began in September 2002.

Design changes following consent

The T5 decision endorsed almost all of the proposals that had been presented by BAA to the public inquiry, in particular

- the scale, design and architectural quality of the main terminal building (T5A)
- the principle of confining the development to the east of the A3044, despite the higher densities that this entailed for the principal T5 site
- the stepped profile of the main buildings in the landside areas up to T5A.

Nevertheless, the secretary of state's decision had a number of impacts. In particular, reductions in maximum height limits for the principal buildings within the landside area were imposed in order to reduce the scale of the development and to reveal more of T5A. It soon became clear that these matters gave rise to a number of consequences that could best be resolved by revising parts of the proposals, particularly those for the landside area.

In August 2002, BAA submitted a planning application to vary a number of the conditions imposed by the consent. An application to the London Borough of Hillingdon sought permission to

- change the form of the T5A roof from an undulating waveform to a simple single waveform and deletion of the forecourt canopy
- change the maximum dimensions of the principal buildings within the landside area
- change the external appearance of the principal buildings within the landside area.

The main effects of these led to the following scheme changes from the approved plan.

- The creation of a pedestrian interchange plaza between T5A and the main multi-storey car park, linking the various modes of public transport and providing an open-air space in front of T5A (Fig. 5).
- Landside, the plan approved by the secretary of state had the first 'row' of buildings comprising a car park, a hotel and an office building. Instead, the first row of buildings closest to T5A (separated by the interchange plaza) would now comprise a 4000-space

multi-storey car park (MSCP1) with an integrated departures forecourt at its top level and a combined bus and coach station and arrivals forecourt at ground level.
- The second and third 'rows' of buildings approved by the secretary of state for car parks were now instead safeguarded for the development of a hotel, together with a second multi-storey car park (MSCP2). The idea of a landside office building was dropped due to the size constraints of the site.
- The overall number of car parking spaces was reduced from 8660 spaces in the approved landside proposals to 4600 (4000 in car park MSCP1 and 600 in MSCP2).

The revised proposals maintained the key elements of the inspector's recommendations, such as the elegant and distinctive floating roof and retention of the stepped approach to the principal landside buildings, and also addressed other concerns.

The creation of the interchange plaza led not only to improvements in the passenger experience, way-finding and interchange; it also opened up views to reveal virtually the entire façade of T5A. The overall bulk of the buildings was reduced and the reduction in the scale of the development created more space for landscaping, which enhanced the landside area and improved the setting of the T5 proposals.

The planning application was granted consent by the London Borough of Hillingdon on 27 January 2003. Thereafter, the many conditions attached to both the main T5 application and the supporting applications were discharged on a regular basis in order to meet the construction programme for opening day.

Conclusion

Gaining permission for T5 was an epic task, demanding time, resources and effort unprecedented in the UK planning system. Because of this, T5 has opened years late. Heathrow's recent passenger-handling difficulties are an indication not just of how badly T5 was needed, but also how important it is that in future the scrutiny of major infrastructure projects like T5 should take place within a decision-making framework that fully recognises their significance to the health of UK plc.

Acknowledgements

The author thanks BAA Ltd for giving permission to publish this paper and is grateful for the help provided by former T5 colleagues Paul Fairbairn (then general manager of planning), Liz Southern (then general manager of development) and Tim Norwood (then planning manager).

Fig. 5. One of the main design changes following planning consent in November 2001 was the creation of a pedestrian interchange plaza between the main multi-storey car park and T5A, linking the various modes of public transport and providing an open-air space in front of T5A

References

1. DEPARTMENT FOR TRANSPORT. *Airports Policy White Paper.* The Stationery Office, London, 1985, Cmnd. 9542.
2. DEPARTMENT FOR COMMUNITIES AND LOCAL GOVERNMENT. *Planning for a Sustainable Future White Paper.* The Stationery Office, London, 2007, Cm. 7120.
3. DEPARTMENT FOR TRADE. *Airports Policy White Paper.* The Stationery Office, London, 1978, Cmnd. 7084.
4. DEPARTMENT FOR TRANSPORT. *The Future of Air Transport White Paper.* The Stationery Office, London, 2003, Cm. 6046.
5. NOTT J. Secretary of State for Trade, Statement to House of Commons. *Hansard*, 17 December 1979, cols 35–55.
6. BALFOUR CONSULTING ENGINEERS. *Mogden Sludge Treatment and the Clearance of Perry Oaks Site.* BCE, 1986.
7. HOWARD HUMPHREYS & PARTNERS. *Heathrow Surface-Access Study Report.* HHP, 1987.
8. VANDERMEER R. *Main Report to the Secretary of State for the Environment, Transport and the Regions on The Heathrow Terminal Five and Associated Public Inquiries.* Two Volumes. Department for Transport, 2001.
9. PALMER D and WILBRAHAM P. Heathrow Terminal 5: twin rivers diversion. *Proceedings of the Institution of Civil Engineers, Civil Engineering*, 2008, 161, Special Issue—Heathrow Airport Terminal 5, May, 25–29.
10. NEVE E. C. *Decision letter on behalf of the Secretary of State for Transport, Local Government and the Regions on applications, schemes and Orders relating to a proposed fifth terminal at Heathrow Airport,* 20 November 2001.

What do you think?

If you would like to comment on this paper, please email up to 200 words to the editor at journals@ice.org.uk.

If you would like to write a paper of 2000 to 3500 words about your own experience in this or any related area of civil engineering, the editor will be happy to provide any help or advice you need.

Proceedings of ICE

Civil Engineering 161 May 2008
Pages 10–15 Paper 700061

doi: 10.1680/cien.2007.161.5.10

Keywords
airports; contracts & law; risk and
probability analysis

Andrew Wolstenholme
BSc, CEng, MICE

is capital projects director at BAA

Ian Fugeman
BEng, CEng, FICE

was formerly BAA head of T5 rail
and tunnels

Fiona Hammond

was formerly construction lawyer
at BAA

Heathrow Terminal 5: delivery strategy

The success of the five-year construction phase of Terminal 5 at London's Heathrow airport was dependent of putting into effect the principles of a unique form of contract, called the Terminal 5 agreement. The £4·3 billion scheme required the client to lead in areas that were typically the traditional domain of suppliers or contractor organisations, resulting in novel methods and relationships. From the early stages it was recognised that the project had to be delivered differently to the norm if it was to achieve its desired objectives. This resulted in the phrase that all those involved in the delivery of the T5 were part of 'history in the making'.

The Terminal 5 (T5) project at Heathrow airport in west London represented not only an unprecedented and massive expansion of the world's busiest international airport but also a risk to the very viability of BAA, the owner and client body. Success would be measured against achieving all the four targets of safety, time, cost and quality. The risk of failure against any one of the targets was sufficient to demand that the delivery strategy had to be different.

The UK construction industry in the mid-1990s could not alone be relied upon to meet the aspirations of the client and thus manage the risks the project presented. 'Doing it differently' soon became a conscious theme of the delivery strategy to bring all parties together with a common commitment that all four targets were to be achieved. This was not negotiable.

By any measure, safety was an absolute requirement, owing to morals, lost time or reputation. The timely delivery of the facility at Heathrow was essential if significant overcrowd-

Fig. 1. The £4·3 billion Terminal 5 project was a massive financial risk to client BAA: completion on time within budget was thus vital

Institution of Civil Engineers

CIVIL **ENGINEERING**

ing owing to the steady increase in air travel was to be avoided. A companion paper[1] provides background to the planning process and details of the prolonged inquiry. Originally the inquiry process was anticipated to last some 18 months, with construction commencing in 1997 and the new terminal facility opening in 2002. As a consequence of the prolonged planning inquiry, the start date for construction became August 2002.

It was critical that the delivery team achieved its target of a five-year construction programme followed by six months of operational trials leading up to the opening on 27 March 2008 (Fig. 1). The risk of cost overrun was perhaps more critical than other projects. Infrastructure projects of this scale often have recourse to the public sector purse if cost escalation occurs, but this was not an option.

When the project started, BAA was a leading company on the FTSE 100 share index and subject to the intense scrutiny by its shareholders and city analysts as well as the regulator, in respect of the company's monopoly of London's major airports: Heathrow, Gatwick and Stansted. A five-year review by the regulator determines the value of landing charges and these had just been set prior to the decision by BAA in March 2002 to proceed with the construction phase. The regulator also set milestones for the construction phase which, if not achieved, would result in the freezing of the agreed phased increases.

When the decision was made to proceed in March 2002, BAA set the budget at £4·3 billion, which was equivalent to approximately two-thirds of its capital value. By most standards this was a bold decision. In addition, BAA had by that time already spent some £540 million.

The fourth target of quality manifests itself in many ways, from the cost of rework and failure to achieve an integrated and functioning system to the very subjective customer experience or the form and quality of the built environment. Clearly 'customer' includes all parties, be it those who work within the facility to the traveller who enjoys a transitory experience forming a part of their journey.

Thus, when BAA was contemplating the challenges presented by the immense T5 programme, it realised that it needed to look at the project in a different way from projects it had already delivered within its estate. The key project strategy drivers were identified at an early stage. These underpinned the approach adopted to deliver T5, which in turn determined the contractual arrangements.

The T5 agreement

Understanding the challenges presented was informed by learning from experience elsewhere, including the Heathrow Express development, and taking into account the pressure for industry reform coming from the government (in the form of the Latham report,[2] the Egan report[3] and the Housing Grants Construction and Regeneration Act), from clients and contractors and also, BAA was to learn, from the insurance market. These lessons led to a clear determination of the project execution principles—the principles that would apply to the delivery of T5 as a programme—and, from these, the commercial principles were derived. These were explicitly designed to work with and underpin the execution principles. The totality of these together with a clear statement of behavioural, cultural and organisational expectations created the legal framework which is the 'T5 agreement'.

Background

In order to understand the T5 agreement, it is important to understand the issues that faced BAA in the early- to mid-1990s. The first major issue lay in the scale of T5 as a £4·3 billion project. BAA itself had not before undertaken projects on such a scale, nor previously integrated assets into its existing business on this scale. Secondly BAA came to the realisation that the industry's experience of projects on such a large scale was intermittent. Equally significant was the consideration of T5 as a business proposition, which is no doubt true of all projects. Set in the context of a very challenging budget and programme, it was clear that T5 needed to be thought about in a different way.

In terms of delivery, there were the usual risks including the unknown, for example future technological developments. Similarly, the environment in which T5 would be delivered (i.e. the legal and governance requirements), was also evolving. These challenges are true of many projects. The requirements of British Airways, the single tenant, changes rapidly with the dynamic nature of the airline industry and in response to the opportunities presented by advances in technology. The significant escalation in the demand for measures to counter acts of terrorism subsequent to attack on the World Trade Center in New York in September 2001, and further developments in the summer of 2006 in the nature of the threat, demonstrates the evolving nature of the form and functionality of the facility. The magnitude and timeframes of T5 meant, however, that these risks had to be implicitly understood and addressed in the agreement. The conclusion reached was that BAA needed an agreement that could enable the project teams to be very flexible in its approach.

On the basis of its research into projects of over £1 billion, and its own experiences at Heathrow Express and elsewhere, BAA had learned that processes and organisation needed to be designed to expose and manage risk, to promote and motivate success and opportunity and to address the behaviours required in all of the key relationships. BAA's thinking about contracting had thus evolved over time from the traditional transactional approach to a more relational approach. This required a change of mindset, and this is reflected in the T5 agreement.

Principles

Having carried out research into the issues facing BAA at T5, the project team then devised a set of simple but compelling execution principles. The T5 agreement is fundamentally built around three simply stated success themes

- do what you are doing well and do it better
- understand 'how' you will deliver as well as 'what' you need to do—this means addressing organisational development as well as technical development
- continually work on the relationships including those that are inward-facing/ inside the project and those that are outward-facing.

When these concepts are translated into the arena of risk and opportunity, then it became clear that BAA needed to be sure that it could identify and manage risks. This meant

- employ and thoroughly apply good basic management practices
- expose and manage risks rather than seek to transfer or bury them
- create a watchful and responsive culture, because the biggest risks are the ones which are not or cannot be identified
- remove commercial disincentives to manage risk and instead create the right commercial environment to enable the achievement of objectives via well-managed risk.

It was also clear that in order to achieve the required changes, a culture needed to be created in which teams could actively promote and pursue opportunities. In the T5 context this meant creating a culture in which people were encouraged to

- seek out, capture and exploit the best practices of others
- remove the barriers and inhibitors to doing things differently
- stimulate and support good ideas
- leverage the commercial incentive to perform exceptionally.

All of this needed to be underpinned by organisational development, as understanding the need for organisational changes could lead to changes in the way that teams and companies behave. In turn that meant empowering leaders; creating integrated teams who would work to common agendas based on co-operative relationships; and incentivising people to solve problems together and act on learning, rather than allocating blame or exploiting the failure or difficulties of others for commercial advantage.

The commercial principles were then designed to link directly to and underpin the execution principles. As most of the effort was to be directed at making things work, people would be rewarded for working at being successful. The commercial arrangements would remain fair and balanced; norms and performance measurement would be used to demonstrate the best commercial deal, and predictability could be achieved by remaining flexible to adapt to circumstances if solutions were found to be insufficient to meet their targets.

It also required that BAA would be able to understand its costs fully. A key principle therefore was total cost transparency, with open access to information on value and waste—not just the numbers—and then planning to decide which costs were of value, by

- incentivising the use of directly employed resources
- pre-planning risk responses and agreeing accountabilities before work started
- examining costs, and then challenging and planning to avoid spending on wasteful activities
- measuring progress and performance
- incentivising exceptional performance by pre-planning for exceptional performance, and sharing the benefits.

It was also recognised at the outset that the commercial arrangements needed to be fair and balanced. This meant paying for proper performance, that is works done to plan and risks managed as agreed, using appropriate re-imbursement routes and early payment of shared benefits. Underlying that were balanced, legally sound safety nets, underpinned by appropriate insurances. The aim was to reduce the incentives for poor time and cost control; thus there would be no profit on below-par performance and the client retained rights of termination on failure to deliver performance.

In essence, BAA's research and learning helped it to understand that to deliver T5 safely it

- needed partners not suppliers
- had to be ruthlessly realistic about the risks it faced and ensure they were properly managed
- had to create a culture where opportunity was pursued
- needed everyone on the programme to be successful in managing risks
- had to recognise that being successful meant doing something different
- had to create an environment in which people and companies felt safe to do something different

- had to create an environment that enabled people to perform
- had to create an environment in which people were motivated to deliver exceptional performance.

These principles are the ones from which the T5 agreement is derived and underpin the whole tenor and approach of the agreement. In understanding the T5 agreement, it is fundamental to recognise it is designed explicitly to address these issues, to enable people to work together to deliver exceptional performance and to achieve success together.

Integrated project teams

The T5 delivery strategy centred on the concept of the project being composed of a series of products delivered by integrated teams, comprising a fully integrated supply chain in which BAA itself took a proactive leadership role.

Teams were not formed conventionally, that is not by company or discipline, but were assembled around 'customer products'. At T5 the teams were to be made up of individuals identified as having the right skill sets for the activity in hand, irrespective of who employed them. It was envisioned as a 'virtual' organisation, integrated at several levels and rationalising skills from across consultants, contractors and suppliers, including necessary BAA skills, and integrating the client development team in delivering solutions. With hindsight it can be seen that this was difficult to achieve consistently.

There were many project teams where the integration was successfully achieved, however, as demonstrated by the manner in which they managed unforeseen events in line with the project objectives rather than enhancing the financial position of an individual company.

All risk on client

A much-heralded concept is that, at T5, BAA held all the risk all of the time. To understand this concept, especially as it was used to underpin the commercial strategy for the T5 agreement, it is important to distinguish risk, in the sense of the potential for harm or opportunity, and liability, in the sense of who pays when things go wrong. The word 'risk' is commonly used as a catch-all, and the phrase 'risk transfer' is often used in the context of contracting with a supply chain. The common use of the language of risk transfer does not, overtly, make that necessary distinction between the risk of harm occurring to a party and liability of another to pay for it. The T5 agreement explicitly does so.

What BAA had realised was that, although it might be able to transfer the liability—the obligation to pay—when a risk materialises, it could never transfer the risk itself, in the sense that it (BAA) would always be the entity that suffered the harm and that frequently recovery of a sum

of money from a third party would unlikely repair that harm to any meaningful extent. In any event, BAA had recognised that some harm cannot be quantified, at least for purposes of establishing contractual liability.

The notion that risk is a fact of life, and that contracts cannot change facts of life, is therefore a powerful idea underpinning the T5 agreement. The T5 agreement saw the main risk to be a failure of BAA to meet its objectives, the consequences of which cannot be meaningfully transferred to the supply chain. It was therefore concluded that, except in very small measure to ensure propriety in the supply chain, it was not the object of the contract to do so.

In an environment where suppliers were providing individuals to work in integrated teams alongside individuals from other suppliers (the virtual company), and where those teams were to be led by BAA staff or individuals from different suppliers, the notion that risk can be transferred to any given supplier is nonsensical. In that context, expecting suppliers to price risk would also be commercially meaningless and could not be said to represent value for money as it would be impossible to hold an individual supplier to account.

Shared liabilities

Under the T5 commercial strategy, the liability (as distinct from the risks) of the suppliers and BAA was to be shared on a strict no-fault basis. For suppliers, this liability was capped (with some exceptions) by the amount of the available incentive fund. Clearly this would not be sufficient to cover the consequences of all the potential risks and BAA therefore took out employer-controlled insurance programmes to address major or catastrophic risks on a project-wide basis covering the whole of the supply chain. These insurances were construction all risks, third-party liability and, a first, project professional indemnity insurance. This was also bought wherever possible on a no-fault basis so that legal liability was not an issue to be determined prior to the policy responding.

First-tier suppliers were therefore paid actual costs (a defined term in the contract and defined against an agreed cost model) and a fair, agreed, ring-fenced profit. The suppliers' contractual liability would be a predetermined share (without proof of fault) of the financial consequences of any risk that occurred (risk in this context included defects and non performance). This liability was capped by the amount of the incentive fund except for certain specified indemnified and insured risks.

Under the exceptions, legal liability was, in theory, unlimited. They were however, with the exception of employer liability insurance, insured under the insurances taken out by BAA. Those insurances were for sums far greater (because they were taking account of the totality of the project) than would have been contemplated by any given contractor contemplating insurance for its own defined piece of work. One area in which liability was clearly transferred to suppliers was in the area of the insurance excesses. In the event of a claim, these were shared among all members of the relevant team or teams (including BAA) in pre-agreed proportions and elements of them were payable from the built-up incentive funds. If there was no incentive fund or insufficient fund available, then the excesses fell wholly to the members of the relevant team.

It can therefore be seen that the incentive fund was a key part of the commercial strategy. Suppliers had had no opportunity to price risks or liability and the incentive fund was therefore their only opportunity to improve on the margins that were represented by the ring-fenced profit arrangements.

The delivery and commercial strategies were therefore designed in combination to change the focus of the contract to management, rather than transfer of risk. This was to ensure that the project teams' focus was on managing the cause up front and not the effect (in terms of claims and counter claims) after the event.

Cultural commitment

Unusually for a contract, the T5 agreement also addressed the cultural requirements that it believed were a prerequisite to successful delivery of the project. This was the third limb of the arrangement. BAA accepted that in strict legal terms it is difficult to enforce such provisions and aspirations. These issues were, however, sufficiently important that, in BAA's view, they needed to be expressly dealt with within the contractual terms. The contract therefore explicitly requires individual and firms to be aware of and focus on partnering, trust and cooperation, and being seen to do what they say.

Procurement

Approximately 75% of the £4·3 billion project cost was procured via the T5 agreement in contract with its 80 first-tier suppliers (Table 1). The remaining value was associated with works being procured or liabilities incurred via a variety of forms of contract appropriate to the party involved, for example Thames Water, Highways Agency, London Underground, Network Rail, Heathrow Airport Ltd. A BAA version of the NEC Engineering and Construction Contract was the only recommended form for the thousands of second-tier contracts in the supply chain.

BAA also used various NEC contracts—particularly the Professional Services Contract—for around 150 direct relationships with consultants (Table 2) and other suppliers, representing approximately 10% of total project cost.

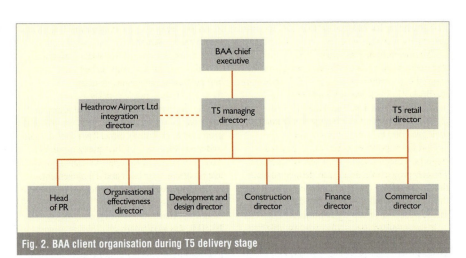

Fig. 2. BAA client organisation during T5 delivery stage

Table 1. Top 30 by value (all over £20 million) of the 80 first-tier suppliers—all were engaged under the bespoke T5 agreement

Rank	Name	Service
1	Laing O'Rourke Civil Engineering Ltd	General civil engineering and concrete structures
2	AMEC Group Ltd (M&E)	Mechanical and electrical installation
3	Rowen Structures Ltd	Supply and erection: structural steelwork
4	Vanderlande Industries UK	Design and supply: baggage transfer systems
5	AMEC Group Ltd (Civils)	Aircraft pavements
6	Mace Solutions Ltd	Fit-out contractor
7	Morgan Vinci JV	Tunnelling contractor
8	Crown House	Mechanical and electrical installation
9	NTL Group Ltd	Communications and controls infrastructure
10	Balfour Beatty Rail Projects Ltd	Rail and tunnel systems
11	Balfour Kilpatrick Ltd	Mechanical and electrical installation
12	Honeywell	Fire alarm and control systems
13	Hotchkiss Ductwork Ltd	Ventilation ductwork
14	Mansell plc	Fit-out contractor
15	Balfour Beatty Construction Ltd	Fit-out of rail and tracked transit system stations
16	Lindner Schmidlin Facades	External cladding
17	Foster Yeoman Ltd	Supply of aggregates
18	Hathaway Roofing Ltd	Roofing
19	Schindler Ltd	Lifts
20	Ultra Electronics	Communication systems
21	Bombardier Transportation (Holdings) USA inc	Design and supply of the automated people-mover system
22	Seele, Austria GmbH	Internal glazed screens
23	Kone Plc	Lifts and escalators
24	Menzies Aviation Group (UK) Ltd	Logistics and transportation
25	Lindner Ceilings Floor Partitions PLC	Ceilings
26	ThyssenKrupp Airport Systems	Aircraft docking bridges
27	Vetter UK	Flooring
28	Blue Circle Industries Ltd	Supply of cement
29	Permasteelisa	Internal glass walling
30	Elliott Group	Supply of temporary building

A much-heralded concept is that, at T5, BAA held all the risk all of the time

Table 2. Main professional services suppliers (value greater than £5 million)—all were engaged under the NEC Professional Services Contract

Rank	Name
1	Richard Rogers and Partners
2	Pascal and Watson
3	HOK
4	Mott MacDonald
5	ARUP
6	EC Harris
7	Turner and Townsend

Programme organisation

The organisational structure of the client and main delivery management teams naturally evolved and changed over time. Fig. 2 shows the BAA T5 directorate applicable during the delivery phase and Fig. 3 identifies the 16 projects comprising the T5 programme and their grouping.

First-tier suppliers worked in a variety of projects, which in turn comprised sub-projects. The sub-project execution plan defined scope, time, deliverables and the team membership and their roles, and was a contractual document. There were 147 separate sub-projects.

Project insurance

The contractual arrangements and risk-sharing model of the T5 project were complementary to BAA's pre-existing philosophy of employer-controlled insurance arrangements for major projects. Such an approach, as opposed to a contractor-controlled insurance programme, had a proved track record within BAA for providing the following benefits

■ cost effective
■ consistency and certainty of cover
■ control over cover and claims
■ ownership of insurer relationships
■ ease of management
■ ability to interface with operational covers.

It was also important that the insurance arrangements would support and encourage the behaviours that BAA was striving to achieve on T5. The was particularly relevant to professional indemnity insurance, where traditionally significant legal costs and programme delays can result from a dispute over a breach of professional duty.

The recognition of the risks to which BAA and the T5 project were exposed informed both the types of insurance cover that were needed (i.e. construction all risks, public liability, marine cargo and professional indemnity insurance) as well as those which were not (e.g. delay in start-up).

The insurance market's standard forms of cover were deemed adequate to cover all but the professional indemnity risks. BAA thus decided that a bespoke form of 'first party, no-fault' cover was required to support the integrated team-working approach and T5 agreement.

An insurance broker was engaged to liaise with the insurance market and explain the unique approach to risk being adopted on T5. This was particularly challenging against the backdrop of the October 1994 Heathrow Express tunnel collapse and involved an extensive programme of face-to-face meetings, site visits and the establishment of a dedicated 'T5 room' at the broker's office into which technical data, plans and artists impressions were placed for easy access by insurance underwriters. BAA also committed to comply with the joint of code of practice being developed by the Association of British Insurers and the British Tunnelling Society in advance of its publication.

It was imperative to keep the insurers convinced on the validity of the delivery strategy and of BAA's commitment, and they were provided with the ongoing evidence that the approach was working. To this end, by mutual agreement, the insurers appointed an engineer who was located on site and had unhindered access to the site and personnel, to act as the insurers' eyes and ears.

Industrial relations and employment

While the responsibility for good industrial relations and fair employment policy is between the employer and the employee and their representative organisation, the consequence of a poor relationship is felt directly by the client body.

The risk that individual suppliers would adopt differing terms of employment and possibly result in unstable industrial or employee relations led BAA to formulate the T5 industrial relations policy and the T5 employment policy. At T5 BAA entered into contract with 80 first-tier suppliers individually but under identical terms in the form of the T5 agreement. Initially some 65 suppliers signed up in 1999 and worked with BAA during the development stage leading up to start of construction activity. The T5 industrial relations policy was incorporated in the T5 agreement and was thus contractually binding.

The policy required the suppliers to

■ ensure every construction operative is directly employed
■ provide transparency and alignment for operative pay
■ ensure any bonus arrangement to have a measurable basis
■ operate within the framework of national agreements and thus employ individuals under one of the four recognised national agreements accepted on the project (Fig. 4).

The T5 programme was nominated as a 'designated' project under the scope of the UK construction industry's major projects agreement (http://www.mpaforum.org.uk/). Alternative arrangements were agreed on an occasional basis but only when they did not present a risk of industrial relations instability, impact on BAA's reputation or increase the cost of labour across the programme. BAA, the suppliers' managers and the representatives of the trade unions met frequently to ensure the fair application of the policy and that any issues were identified and resolved.

The T5 employment policy provided practical advice and standards on all aspects of employment of construction operatives including

■ recruitment, covering use of a common recruitment database, interview and application process, establishing 'right to work in the UK' in accordance with the Asylum and Immigration Act 1966, security screening and verification of competency and skills including a minimum period of experience in the construction industry
■ during employment to ensure that employees were inducted and properly trained, possessed nationally accredited skills certification or the Construction Skills Certification Scheme (CSCS) health and safety test, completion of a health questionnaire or in safety-critical roles to be seen on site by an occupational health nurse
■ while on site all personnel were to comply

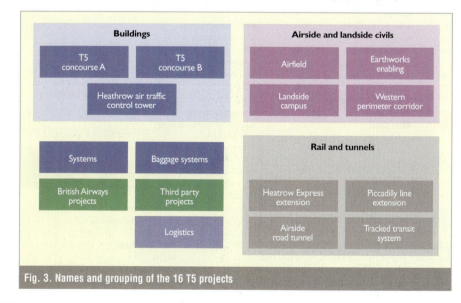

Fig. 3. Names and grouping of the 16 T5 projects

Buildings
T5 concourse A
T5 concourse B
Heathrow air traffic control tower

Systems
Baggage systems
British Airways projects
Third party projects
Logistics

Airside and landside civils
Airfield
Earthworks enabling
Landside campus
Western perimeter corridor

Rail and tunnels
Heathrow Express extension
Piccadilly line extension
Airside road tunnel
Tracked transit system

with the policies regarding drugs and alcohol, health and safety, and other site rules
- the employer to comply with a no-poaching policy, to have proper procedures for managing grievances and disciplinary issues, when employing individuals from abroad to ensure compliance with work-permit requirements and, for non-English speaking personnel, specific requirements to ensure the individual understands the training and the task
- pay, reward for productivity, resignation, dismissal, end of contract and redundancy
- assurance as well as the requirement for the first-tier suppliers to audit their sub-contractors to ensure compliance to both policies.

In over five years of construction activity there were six days of industrial action contained within specific parts of the works, and which had no impact on the overall programme. The commercial implications of any changes in agreements were managed between BAA and the individual first-tier suppliers.

Commitment to people

BAA recognised that it had to lead on the application of the processes to achieve the cultural change which was implicit in the delivery strategy and expressed in the T5 agreement.

The introduction of a safety initiative known as the 'incident and injury free' programme, required a cultural change to enable each individual to recognise that there is no acceptable level of accidents. BAA led by mobilising all personnel, staff and operatives from all participating organisations, from managing director level down to commit to the belief in the possibility of a programme that was incident and injury free. The initiative, which is explained in detail in a companion paper,[4] was supported by poster campaigns to encourage individuals to take personal responsibility for their actions.

BAA's commitment to respect the individual was expressed in the establishment on site from day one of a very high standard of office accommodation and welfare facilities. A constrained site and its locality meant that travelling requirements for all personnel had to be addressed and resulted in off-site controlled parking and extensive bus facilities.

The achievements of individual team milestones were celebrated and supported by an award scheme, ranging from a regular £5000 team award down to £25 spot awards for exceptional performance or initiatives. Communications of the expected behaviours as well as news of the project were published monthly in an extremely effective site newspaper. For office-based personnel, a project intranet site utilised the single IT environment

which supported all project communication and production of drawings and documentation.

The response to individual needs was an essential part of BAA's policy to manage risk and thus honoured the T5 agreement in support of the delivery strategy.

A continuous theme or branding supported the various specific initiatives and sought to achieve pride in the project by association with other iconic structures. Thus all involved could one day say 'I was proud to have built T5' and was part of 'history in the making' (Fig. 5).

Fig. 4. The industrial relations policy incorporated in the T5 agreement required that all construction operatives were directly employed under one of four recognised national agreements

History in the making.

Tip-top: final section of the 66,000 sq.metre, 17,500 tonne, 'Concourse A' roof is raised into position – Spring 2005.

One day you'll be proud to say,
"I built T5."

The world's most successful airport development

Fig. 5. This 'History in the making' poster was one of a number of initiatives that sought to achieve worker pride in the project

References

1. PELLMAN R. Heathrow Terminal 5: gaining permission. *Proceedings of the Institution of Civil Engineers, Civil Engineering*, 2008, **161**, Special issue—Heathrow Airport Terminal 5, May, 4–9.
2. LATHAM M. *Constructing the Team—The Final Report of the Government/Industry Review of Procurement Arrangements in the UK Construction Industry.* HMSO, London, 1994.
3. EGAN J. *Rethinking Construction.* Department for Trade and Industry, London, 1998.
4. EVANS M. Heathrow Terminal 5: health and safety leadership. *Proceedings of the Institution of Civil Engineers, Civil Engineering*, 2008, **161** Special issue—Heathrow Airport Terminal 5, May, 16–20.

What do you think?

If you would like to comment on this paper, please email up to 200 words to the editor at journals@ice.org.uk.

If you would like to write a paper of 2000 to 3500 words about your own experience in this or any related area of civil engineering, the editor will be happy to provide any help or advice you need.

Proceedings of ICE

Civil Engineering 161 May 2008
Pages 16–20 Paper 700042

doi: 10.1680/cien.2007.161.5.16

Keywords
airports; health & safety

Mike Evans
BEng, CEng, MICE, MIOSH

is head of health and safety at
Terminal 5

Heathrow Terminal 5: health and safety leadership

Heathrow airport's Terminal 5 project has gained widespread recognition for a safety performance some four times better than industry norms and for setting new benchmarks for occupational health and safety. The sheer size of the £4·3 billion project, its structural and managerial complexities and the 60 000 people that have worked on site have presented significant challenges. As this paper describes, safety leadership—as distinct from safety management—coupled with real engagement of, and respect for, all concerned has led to cultural change. This has not only reduced the number and severity of injuries but has also resulted in improved worker health, satisfaction, morale and performance.

From 2002 to 2007, the £4·3 billion Terminal 5 (T5) project at London's Heathrow airport was one of the largest construction projects in Europe (Fig. 1). With a peak population of over 8000, it has brought together a total staff and workforce of more than 60 000 people.

From the outset it was recognised that if T5 was to achieve only average UK construction industry safety performance levels then several people would die, hundreds would suffer major disabling injuries and thousands would be hurt in some way. Client BAA deemed this unacceptable from both a moral and business perspective. Additionally, the challenge of attracting, retaining and motivating the necessary workforce was seen as requiring more than just good rates of pay. Excellent health, safety and welfare provisions, training and up-skilling facilities, assistance with securing local accommodation and ease of transport to site were all seen as key factors in making T5 a great place to work and thus creating a stable industrial relations environment.

Health and safety was therefore seen as a critical element to the success of T5 and one of several areas where 'something different' was needed to raise industry performance levels.

Health and safety objectives

The overall objectives for BAA on T5 were to deliver a quality product, on time, within budget and safely—traditional construction project priorities that always compete with one another and are rarely all satisfied at the same time. T5 directed its efforts toward treating safety differently—as a 'core value' for everyone involved and not simply as one of those competing priorities. The intention was to create a 'people-focused' regime rather than one based solely on statutory compliance. The vision was to create a culture intolerant of incidents, injuries, unsafe behaviours and poor attitudes to health and safety—in short, to journey towards making T5 incident and injury free. The health and safety objectives were therefore

- to elevate everyone's awareness of safety by making it personal, relevant and important to every person on the project to create a fundamental shift in attitudes and culture
- to position safety as a core value rather than one of a list of competing priorities
- to create a vision and expectation for future safety performance rather than focus on historical statistics
- to set new industry standards for health, safety and environmental performance.

In the creation of this cultural change programme, great care was taken to ensure that the 'people safety' agenda in no way detracted from the 'process safety' issues. The

Fig. 1. Over 60 000 people worked on the 126 ha, £4·3 billion Heathrow T5 project from 2002 to 2007, making it one of the largest construction projects in Europe—but also four times safer than average

structural safety and integrity of permanent and temporary works and construction methods; minimisation of risks to the surrounding operational airport; statutory compliance; general risk management and health and safety processes all remained key issues as the cultural change programme developed.

Safety in design

Safety considerations were inherent from the earliest stages of design with specialist input provided by the T5 planning supervisor when the Construction (Design and Management) (CDM) Regulations first came into force in 1994. The planning supervisor became an integral part of the project management team from concept design stage and, most importantly, was co-located with the integrated design teams throughout the design process.

Each design team nominated a member as their 'CDM co-ordinator' (a term now adopted in CDM 2007 in place of 'planning supervisor'). This encouraged each team to be self-accounting for their CDM obligations, provided a single point of contact for the planning supervisor and, through regular meetings, enabled sharing of information and commonality of approach towards risk mitigation.

The planning supervisor also developed a 'red list' of products and processes that were known to be harmful to people or the environment. Specification of such items necessitated approval via a justification report, stating why it had not been possible to find a safer or healthier alternative. This approach is now growing in popularity and is endorsed by the UK Health and Safety Executive (HSE).

A real drive to ensure that construction methodologies were developed in conjunction with the design, via contractor representation

within all of the integrated design teams, was realised. Opportunities for prefabrication and modularisation were actively pursued and off-site trial assemblies were also carried out on major building components to test the assembly processes and learn lessons off-site and off the critical path. Technical innovation and process improvements were always encouraged but only adopted where an overall benefit to the project could be clearly identified against trusted technology, with safety considerations always to the fore in such deliberations.

Principal contractor appointments

The HSE was consulted on proposals for T5's unique 'principal contractor' arrangement, where the integrated team approach to project management meant it was difficult to identify a single organisation to act as principal contractor across the entire site. Some elements of T5 were easily segregated and had an identifiable lead contractor best placed to become the principal contractor, for example bored tunnels, rail systems and the rail station. However, the bulk of the civil engineering and building works were not easily segregated and it was therefore proposed that a joint organisation, the 'T5 principal contractor', be established, comprising BAA and three of the main contractors. This arrangement, accepted by the HSE, not only confirmed BAA's hands-on project management role but also firmly established a shared responsibility and ownership for health and safety across the whole site among the major players.

To implement the arrangement, the senior production manager for each major element of work was formally appointed as the 'principal contractor's representative' for that element and its defined geographical area of the site. This became a powerful and highly effective arrangement for the allocation of health and safety

responsibilities across the site as well as adding an element of competition.

Safety in construction

Health and safety management system

Many months were spent planning the project health and safety management arrangements and how the health and safety objectives would be achieved through a high-level principal contractors' forum, which developed a bespoke safety management system for the project. Commonality of procedures, standards and rules were seen as fundamental on a complex, multi-contractor site with shared safety responsibilities.

The initial intent in developing common procedures was to address only key high-risk areas, such as control of service information for critical airport services and fuel mains, and the management of cranes. However, the benefit of sharing best practice amongst the major players soon became apparent and a suite of T5 procedures was developed to support the safety management system. This process served not only to raise standards for everyone but also to foster a good working relationship between all the principal contractors.

The safety management system addressed all of the usual components of a competent safety management system, but with a particular focus on people and communication issues, including

- mandatory competency standards (e.g. Construction Skills Certification Scheme cards)
- site-based training facilities including a health and safety test centre
- formal and informal consultation processes
- an award-winning three-stage induction process
- frequent and high-quality communication processes (e.g. site safety booklets, mandatory daily activity briefings, site newspaper and corporate communications campaigns (Figs 2 and 3))
- comprehensive safety support services and site safety hotline
- high-profile monitoring and feedback process.

The T5 safety management system, the supporting procedures and site rules were all developed jointly by the principal contractors and were published in the client's name as mandatory across all parts of the project, including later third-party projects.

Shared services

A large, dedicated logistics team developed detailed plans for the provision of common services to all contractors on site. These included car parking, buses, induction, security, utilities, roads, site accommodation, catering, welfare, provision of bulk materials, off-site logistics areas, lorry parks and so on.

The whole fabric of the site and site infrastructure was therefore determined and delivered, to an award-winning standard, through the principal contractor and logistics team for the benefit of all. This was seen as a prerequisite to delivering management's commitment to improved health and safety.

Occupational health

Occupational health (OH) was an area identified as particularly important under the 'respect for people' strategy with a real desire to provide an exceptional service on T5. The OH provider appointed was chosen because it offered 'intelligent' and proactive support services designed around the needs of the project and the workforce, rather than a menu of standard OH screening services.

The resulting OH strategy was designed around two key principles, focusing on preventative measures rather than just treating the symptoms

- the effect of health on work—people's health must not jeopardise their ability to work safely
- the effect of work on health—people's work must not jeopardise their health.

Fig. 2. The monthly newspaper The Site was an integral communications element of the health and safety management system

Fig. 3. The 'injury and incident free' (IIF) programme promoted the common goal of everyone going home safely every night rather than simply adhering to safety rules

Key elements of the strategy included

- pre-start screening for everyone working at T5
- full medical assessments for those engaged in 'safety-critical work'
- education programmes aimed not only at the workforce but also at designers, construction managers and supervisors (a new experience for most)
- comprehensive lifestyle medicals to encourage people to think seriously about health issues, both within and outside the work environment
- health surveillance programmes tailored to suit the needs of the many work processes and employers' existing provisions
- treatment and emergency response services through the provision of nursing staff qualified in both OH and accident and emergency disciplines
- drug and alcohol testing in pursuance of a strict policy supported by all major players and trades unions.

The most prevalent health issues identified from both pre-start screening and routine health surveillance were high blood pressure, undiagnosed diabetes and visual defects. Most surprising was the large number of relatively young, apparently fit people with extremely high blood pressure. Despite much concern within the industry on the subject of hand–arm vibration syndrome, very few cases were identified and these were generally in people with a history of the mining industry.

Through the provision of 'people-focused' services and support to management, the OH team won the trust and respect of everyone at T5—employers, workforce and unions alike.

Benefits arising from the approach included not only improved general health of the workforce, and hence improved worker satisfaction, morale and industrial relations, but also decreased risk of ill health causing a major

accident. A direct financial benefit was also realised—the time savings arising from on-site treatments, consultations and reduced ill health more than covered the cost of the service.

Cultural change

Having implemented 'best in class' systems and procedures and provided excellent welfare and OH services, there was recognition that something radically different still needed to be done to achieve the stated safety objectives. The T5 'incident and injury free' (IIF) programme was developed to help achieve exceptional health and safety performance. Originating from the oil and petrochemical industries, an external consultant was engaged to help the project learn from the successes of these industries in radically changing safety culture.

The programme is strongly focused on safety leadership, as opposed to safety management, and seeks to engage and enrol management, supervision and workforce in the common goal of everyone going home safely at the end of every day (Fig. 3). It thus promotes care and concern for the wellbeing of everyone on site rather than simply driving compliance with safety rules. It equalises and empowers everyone in a 'just' safety culture.

The IIF programme is underpinned by an integral approach to safety that addresses

- proven systems and procedures
- the promotion of desired behaviours
- influencing individual attitudes
- the creation of a shared culture that is intolerant of incidents and injuries.

It is based on a model created by Ken Wilber as shown in Fig. 4. The premise of the programme is that the traditional safety inputs on the objective side of the model can be greatly enhanced by focusing more on the subjective areas. Simply put, identifying motivators behind actions at either individual or team level allows the root causes of behaviours to be addressed.

	Subjective	Objective
Individual	**Intention** • Values • Attitude • Commitment • Responsibility • Experience • Mood	**Behaviour** • Plans • Actions • Decisions • Performance
Group	**Culture** • Shared values • Ethics • Morale • Myths and legends • Justice • Fairness • Covenants	**Systems** • Organisational structures • Work processes • Policy and procedures • Economics • Contracts

Fig. 4. The 'injury and incident free' programme is based on this model by Ken Wilber—the premise being that traditional safety inputs on the objective side can be significantly enhanced by focusing more on the subjective areas

As with the safety management system, the IIF programme was developed jointly with the principal contractors; their ownership was vital for this to be more than just another client-imposed initiative. Key elements of the IIF programme were to

- establish and coach an integrated T5 safety leadership team to provide real and visible leadership of health and safety on T5
- establish and coach a number of project safety leadership teams to ensure effective delivery at individual project level
- train a large number of managers and supervisors to deliver IIF training courses
- coach senior managers in how they might best demonstrate their commitment to safety
- take everyone on site through a tailored event to engage them in the IIF programme.

Implementation of the programme was challenging and some significant barriers needed to be overcome, for example the 'macho' culture and the deeply ingrained attitude that accidents are inevitable. With an average of 1000 new people to the project every month, great efforts were made to engage all with the IIF programme as soon as possible in order to avoid dilution of the improving culture. The changing site population (from a few big employers in the early days to a much more disparate workforce latterly) necessitated ongoing review and changes to the IIF delivery mechanisms and communication processes.

To sustain the long-term programme and maintain its focus, significant efforts were made to identify new leaders and champions and to create new opportunities for them to demonstrate their commitment to safety, sometimes through eye-catching and unexpected actions by senior managers.

In August 2005 and October 2007, two fatalities caused some to question whether the IIF programme and all of the other safety efforts were actually working. However, after allowing time for the site to recover from the tragedies, specific focus was placed on rebuilding belief and becoming even more determined in the safety efforts. The safety commitment that was expected of everyone on site is summarised in Fig. 5.

Health and safety performance

A number of indicators of health and safety performance were used on T5, including accident statistics, incident and near-miss reporting, scored safety inspections, formal safety audits and both formal and informal management safety tours. All of these served to give management an indication of how the project was performing against its stated objectives.

However, in a deliberate attempt to shift the focus from reactive outputs (i.e. accident data) to proactive inputs (e.g. levels of training and planning for safety), a balanced scorecard of safety measures was developed, with a 70:30 weighting ratio applied to the scoring of input and output activities respectively. Contractors and projects were scored on a range of topics and the results published as a monthly league table, with low achievers targeted for specific improvement each month. Typical topics covered and scores available are shown in Fig. 6. Varying either the topics or the weighting of the scores available proved a very effective technique for focusing attention on a specific issue and driving continual improvement.

Accident statistics

Figure 7 shows the accident frequency rate per 100 000 h worked for all Reporting of Injury, Diseases and Dangerous Occurrences Regulations (Riddor) reportable accidents. The 12-month rolling average decreased steadily from 0·69 in July 2003 to 0·18 in December 2007. This compares favourably with national averages of approximately 0·53 from HSE statistics (2005/06)—not allowing for known under-reporting—and approximately 1·10 from the last Labour Force survey (2003/04). Fig. 7 also shows the number of new starters every month, averaging over 1000 per month for the project duration.

Figure 8 shows the minor accident frequency rate per 100 000 h worked for all accident book entries. The 12-month rolling

T5 Incident And Injury-Free Programme

What You Can Count On From Management

- If you stop work for any safety reason, we will support you 100%
- If you raise a safety concern, we will take it seriously. It will not go into a black hole.
- If there is an incident, we will carry out the investigation from the point of view of what we can learn, not who do we need to blame.
- We will look for people working safely and tell them that we see they are and that we appreciate it.

What Management Expects of You

- If it is not safe do not do it and do not let your workmates do it either.
- If you see something that is unsafe SPEAK UP immediately, there and then to your supervisor, no matter who – no matter what.
- If you are not sure of something (it does not look right – you do not understand the task or are unsure of how to do the job safely) SPEAK UP AND ASK.

T5

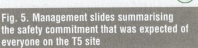

Fig. 5. Management slides summarising the safety commitment that was expected of everyone on the T5 site

H&S Balanced Scorecard

Inputs	Points Available	Outputs	Points Available
Training Managers	16	Reportable Injuries	8
Training Supervisors	14	Minor Accidents	10
Training & Competence Operatives	10	Significant Incidents	12
Meeting Attendance	6		
HS&E Programmes	14		
Bonus Contribution	10		
Total points available - inputs	70	Total points available - outputs	30

Fig. 6. Typical health and safety balanced scorecard, weighted 70% to proactive inputs. A league table of scorecard results for all contractors and projects was published each month

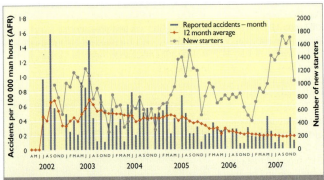

Fig. 7. The 12-month average reportable accident frequency rate (AFR) reduced nearly four-fold over the five-year construction period—from 0·69 to just 0·18 per 100 000 h worked. Despite surging numbers of new starters, this compares with an apparent national average of 0·53

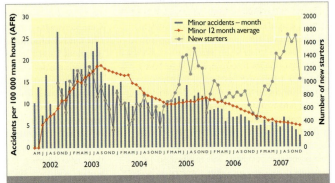

Fig. 8. The 12-month average minor accident frequency rate (AFR) dropped nearly four-fold during construction—from an early peak of 18·6 per 100 000 h worked to just 4·9

average decreased from 18·6 in September 2003 to 4·9 in December 2007. No nationally published data are available for comparison, but the author's experience of the construction industry would indicate that an average accident frequency rate for all minor accidents would be in excess of 20. Again, the number of new starters every month is shown in the figure and there is some evidence of a link between a large influx of new starters in mid-2005 and an increase in the minor accident frequency rate at that time.

The data suggest a safety performance at T5 at least four times better than UK construction industry norms. Further analysis of T5's accident statistics revealed some surprising results, such as the following.

■ Those who had been on site for more than 6 months had proportionally more accidents than those new to the site; it is possible that a degree of complacency crept in.
■ People in the age range 35–45 had proportionally more accidents than those under 25—one possible reason is a 'know-it-all' attitude and older workers less receptive to change.
■ Non-English speakers had marginally fewer accidents than English speakers—possibly due to a culture of compliance among some nationalities or the effectiveness of the measures in place on T5.
■ There were significantly more accidents on Monday mornings and Thursday afternoons than at any other time. It is possible that this was due to a high proportion of the workforce commuting who had thus spent many hours driving to site early on Mondays. Thursday was often the peak production day as workers strived to finish earlier on Friday.
■ There was some evidence of slightly increased accident rates during June and July; this might be linked to decreased supervision levels during holiday periods.

Some of the findings contradict widely held perceptions about the prevalence of accidents within the construction industry. A key lesson is that the design of safety programmes and the targeting of safety resources should be based on a thorough understanding of the particular circumstances of the needs of the organisation, project and workforce, rather than on generalised assumptions or perceived wisdoms.

The T5 'none in a million' safety performance target (i.e. safe hours worked since the last reportable accident) was reached on 18 separate occasions, including three periods of over 2 million hours work without a reportable accident. Hours worked were displayed daily on a large electronic scoreboard at the main site entrance and the 'none in a million' events were

communicated and celebrated as very significant milestones on the IIF journey.

Employee survey results

Whereas statistical data provided one view of safety performance, it was realised that a more valuable perception of how safe the site was could be gained by asking the workforce for their views. The clearest measure of the effect of the IIF programme on culture change came from such surveys, which found

■ 74% of site workers on T5 felt that safety was given more priority than anywhere else they had worked (2005 survey)
■ 96% of all T5 site employees felt that they had a part to play in making T5 a safe place to work (2006 survey)
■ 91% of site workers felt that T5 was the safest place that they had ever worked (2006 survey).

Below the headlines, detailed feedback on workforce perceptions was used to make improvements in the weaker areas. The overview from such feedback was encouraging: workers felt able to talk to their managers and supervisors about safety; managers were seen as mainly responsive; people felt comfortable talking about and reporting accidents and near-misses; workers were happy to stop others working unsafely. Feedback on all of the surveys was published in the site newspaper.

Impact beyond T5

In addition to health and safety successes on T5, there have been wider benefits—BAA is beginning to implement an IIF-type programme within its airport operations and one of T5's major contractors has adopted the IIF programme throughout its business. Other suppliers have adopted key elements of the T5 safety management system.

Senior representatives from the Olympic Delivery Authority have cited T5 as a role model and its success reached the House of Commons when, in May 2007, one MP stated 'the kind of procedures that are in operation at Terminal 5 must be replicated when we build the stadiums for the 2012 Olympics'.

Although only anecdotal, perhaps the most profound examples of T5's safety success can be found from the actions of T5 workers who have left to work elsewhere. One group walked off another job because they were not prepared to accept the conditions and risk of injury in their new contract. It has been reported that one worker was 'ordered' off his new job and back to T5 by his wife because of her concerns for his safety.

All who have seen the real benefits of safety at T5 will understand how such achievements

can be made and hopefully sow the seeds of safety excellence elsewhere.

Conclusion

T5 has gained widespread recognition for its excellent health and safety performance and is seen as having set new benchmarks for the construction industry and beyond. The sheer size of T5 and its structural and managerial complexities have been challenges in their own right. In this context, the delivery of exceptional safety performance, against a background of 1000 new people every month, has been a truly remarkable achievement of which all participants can be justly proud.

It has been made possible through recognition that a systems- and compliance-based approach, while vital in its own right, is not enough to achieve extraordinary performance. Safety leadership—as distinct from safety management—coupled with real engagement of, and respect for, all concerned, has led to cultural change and hence an organisational intolerance of—rather than acceptance of—incidents and injuries. Managers at every level of the organisation have been exposed to the philosophy that the highest level of safety they can achieve is that which they demonstrate they want.

BAA's role as client has been to create an environment where everyone involved—designers, contractors, workforce, unions and the many stakeholders and end-users—has united in the common cause of ensuring people's wellbeing. There have been leaders who shone and inspired, but the main praise should go to all those individuals who were willing to do whatever it took to make sure that they and their colleagues went home safely every day.

There is clear evidence of the cost effectiveness of the health and safety efforts in terms of reduced numbers and severity of injuries, reduced ill health, and hence reduced lost time and improved industrial relations. The relationships established between people at all levels through the IIF safety dialogue have also resulted in improvements in other aspects of performance—quality, cost and programme.

There is a clear message for organisations seeking cultural change: safety is a great place to start.

Proceedings of ICE

Civil Engineering 161 May 2008
Pages 21–24 Paper 700040

doi: 10.1680/cien.2007.161.5.21

Keywords
airports; environment; sustainability

Beverley Lister
BSc, MSc

was head of environment at
Terminal 5 from 2001 to 2006

Heathrow Terminal 5: enhancing environmental sustainability

London Heathrow airport's Terminal 5 project provided an opportunity for owner BAA to set and deliver new standards in environmental sustainability for the construction industry. As this paper describes, opportunities were pursued at each development stage to improve performance and firmly embed environmental awareness and corporate responsibility into decision-making processes. Throughout the design and construction, project teams and suppliers were encouraged to apply innovative techniques and best practice to deliver exemplary environmental performance.

The length of the programme to build the £4·3 billion Terminal 5 (T5) at London's Heathrow airport required a long-term focus to future-proof the development against changing policy, legislation and technological advancements. Indeed, some of the early decisions—such as the use of ammonia chilling within the energy centre, requirements for timber approved by the Forest Stewardship Council (FSC) and avoidance of polyvinyl chloride (PVC)—will deliver environmental sustainability benefits to owner BAA for many years after the terminal's March 2008 opening (Fig. 1).

Design stage

A systematic approach to environmental management was applied to the whole development

Fig. 1. An early focus on environmental sustainability of the T5 project will deliver benefits to owner BAA for many years after its opening

from the beginning of the design stage to identify and manage key issues and monitor performance. Early in the process, an environmental advisory assessment group was created, comprising external environmental experts from a cross-section of the construction industry, including CIRIA, BRE and Forum for the Future. The group was instrumental in advising and steering the project to develop tools that enabled a robust environmental framework from the outset.

The main purpose of the environmental advisory assessment group was to

- advise and participate in the development and application of methodology for environmental assessment
- consider appropriate performance indicators and targets
- advise on proposed legislative and best-practice performance targeting
- review the project design proposals, making recommendations on environmental performance and areas requiring further research.

From this process emerged the key sustainability focus areas, which evolved into the project requirements and subsequent environmental targets, a selection of which are shown in Table 1. Where appropriate, issue strategies such as water demand, waste and materials further defined specific requirements to introduce

In addition, the rainwater harvesting scheme reduces the flood risk of local watercourses by reducing the net runoff from the terminal development. As the downstream surface water network is pumped twice, the scheme also delivers a reduction in energy consumption.

Landscape enhancement

The T5 landscape strategy encompasses four major projects

- M25 widening and spur road
- landside campus
- western perimeter corridor
- Colne Valley.

Colne Valley lies on the western edge of Heathrow airport, bounded to the north by the A4 highway, to the west by the M25 motorway and to the south by a road system called Airport Way (Fig. 4). It comprises a lowland river valley, is designated green belt and forms part of the Colne Valley regional park. Historically, the area has suffered a high degree of disturbance from activities such as mineral extraction and landfill; with many areas also subject to fly tipping and other illegal uses, resulting in a poor-quality landscape. Within this context were a few surviving localised pockets of higher-quality landscape, which reflected former land use and natural landscape elements, in particular the River Colne itself.

Through a number of agreements under section 106 of the Town and Country Planning Act to fulfil public inquiry commitments, BAA and other local landowners have implemented enhancements both to create an approach to T5 and to treat the Colne Valley as an important landscape in its own right.

Design concepts

Whereas the building of T5 and its associated works resulted in a temporary disturbance of the area, a unique opportunity was concurrently in operation to reverse the long-term degradation and enhance the landscape in line with the objectives of the Colne Valley regional park strategy.

In resolving the various and often conflicting requirements of the area, BAA safeguarded the best features, such as the regionally important grassland, and enhanced problem areas to re-establish a viable 'countryside' landscape on the doorstep of the terminal. Landscape improvements, access and recreational opportunities, nature conservation, planning for future users and long-term management formed part of the original plan.

The Colne Valley accommodates the M25 spur as the principal road access to T5. The overall vision for this access route was to create a memorable gateway through a progression of re-established and enhanced agrarian landscapes, graduating to an urban character close to the terminal campus. This is achieved by the creation of a country park landscape, retention and extension of areas of farmland pasture and structured planting on campus.

Implementation

Over the course of a six-year programme from November 2002 to March 2008, the following works were undertaken

- extensive clearance of degraded landscape elements
- creation of landforms from 0·75 million m³ of earth fill
- planting of over 1000 semi-mature trees, 2500 semi-mature shrubs and 50 000 native woodland plants
- planting of 2000 m of native hedgerows to help encourage a wide range of insects and small mammals such as field voles and field mice
- translocation and management of 2·5 ha of regionally important grasslands
- installation of tree sparrow boxes
- creation of new woodlands on road embankments and spoil deposition sites, helping to frame and screen views of the main terminal building
- creation of new and improved access through footpaths and bridleways connecting and bridging a gap that long existed between established rights of way to the north and south of the area
- re-introduction of agriculturally based land and an overall enhanced area of nature conservation.

Furthermore, completed landscape works are subject to a 25-year maintenance commitment post-completion, with the objectives of providing long-term husbandry to promote best establishment of the planting, addressing airport safety considerations (e.g. monitoring bird populations potentially involved in aircraft bird-strikes), maintaining key landscape elements (e.g. hedgerows) and keeping pathways clear and safe.

Waste minimisation and recycling

Many challenges were encountered in the management of wastes generated from the construction, fit-out and future operation of T5. The size and complexity of the project, the number of suppliers and teams involved, and the variety of waste types all required a proactive and innovative approach to construction waste management.

Waste management of the terminal operation was addressed during scheme design. A series of 'waste rooms' facilitate waste re-use and recycling. The modular design enhances building flexibility and deconstruction in terms of future refurbishment. Together with BAA supplier and third-party contractual requirements, these contribute towards delivering BAA's current aspiration of zero waste to landfill by 2020.

During construction, a single waste contractor was appointed to manage and collect all waste from construction areas, site roads, cesspits, offices and canteens of over 150 project teams. Over 87% of all construction waste was recycled, either through off-site sorting and recovery of recycled materials or on-site source segregation. Where possible, designated skips were provided for general, wood, metal and cardboard waste. Waste cables and plasterboard were also segregated for recycling on site; in offices, paper was collected for recycling through a desktop scheme. Waste concrete, topsoil and other aggregates were processed on site under a waste management licence exemption, with a designated team sorting and, where necessary, crushing waste aggregates for on-site re-use as backfill or landscaping material. Even waste arising from sweeping activities on the 25 km circuit of site roads was re-used by processing on site through a series of oscillating screens, filters and belt presses to remove the solids and clays for re-use, with disposal of the cleaner water.

Designers and suppliers were required to seek to reduce waste from the outset. Design teams adopted the principles of standardisation, prefabrication and modularisation, dramatically reducing waste and maximising efficiencies in material and resource consumption. Packaging

Fig. 4. Landscape master plan for the Colne Valley area to the west of the airport

reduction was addressed with fit-out suppliers during the tender stage and with third parties through environmental requirements of their concessionaire submissions. The use of logistics centres for managing deliveries to the terminal allowed for the removal and storage of packaging in a controlled environment, which encouraged supplier participation in take-back schemes for pallets, crates, stillages and cable drums.

Other initiatives helping to reduce waste to landfill included a community paint exchange and a materials take-back scheme that 're-homed' unwanted products and materials elsewhere on site.

Materials 'red list'

The T5 materials strategy was probably ahead of its time when first drafted in October 2000. It was recognised that, as a result of the scope and volume of material specification for T5, there existed the potential to affect significantly both the environment and health of the workforce and building uses. A solution lay in ensuring materials visibility and choice during design. This gave rise to a combined Construction (Design and Management) (CDM) Regulations and environment 'red list', taking account of the commonalities between the two disciplines. Procurement of anything on the list was to be avoided through either design change or the use of alternatives. Environmental components of the red list were

- non-FSC timber
- PVC and unplasticised PVC
- hydrofluorocarbon (HFC) and hydrochlorofluorocarbon refrigerants
- virgin aggregates and cement
- timber preservatives
- formaldehyde
- harmful particulates and volatile organic compounds.

Some of the environmental targets for T5 materials are shown in Table 1. Under the guidance of the environmental advisory assessment group, BAA sought to demonstrate further leadership in the materials field through encouraging the use of innovative, recycled and sustainable materials, as well as minimising the use of non-sustainable or toxic materials.

In order to encourage change and measure success, targets were proposed for the use of red list materials. The targets reflected the relative immaturity in the marketplace for proven alternatives, while striving to drive change within the industry. Everyone responsible for specifying and procuring materials was required to comply with the materials strategy and compliance was audited throughout the design stage and on site. Where materials could not

be designed out or alternatives could not be sourced, a report providing full justification of use had to be submitted and approved by BAA before procurement could proceed.

Awareness campaigns, training and environmental reviews ensured that designers incorporated red list requirements into their designs and specifications; reference to the red list was included in all acquisition plans. However, the multi-tiered nature of the T5 supply chain gave rise to problems in communication and sourcing, and implementation of the red-list strategy became more complex once BAA started procurement.

On several occasions, materials were refused access to site or spotted on site and rejected. Twice, timber products were installed before it was noticed they were accredited under a certification scheme other than the FSC—to the industry it seemed one sustainable timber was the same as any other. Through experience, BAA learnt where the pitfalls were and how to ensure red-list requirements were passed on to suppliers; eventually, change happened.

Use of some alternative materials, such as ethylene-tetrafluoroethylene to replace PVC film, required changes in product design to accommodate the different specification; others, such as formaldehyde-free fibreboard, had to be sourced from Europe. The replacement of HFC with ammonia chilling for the terminal's air-conditioning network required redesign of the energy centre while also challenging BAA's design and safety standards.

There have been some set-backs but, overall, having a materials strategy to challenge the status quo has eliminated numerous harmful materials and working practices that would otherwise have been used in the construction of T5.

Working with the regulators

Close working relationships were fostered with the environmental regulators throughout the design and construction of T5. Integrity and transparency were fundamental in achieving successful partnerships, reinforced by full access to the site at any time, sharing of environmental monitoring data and proactive resolution of issues.

The project teams worked closely with environmental services teams at three local authorities in the management of air quality, dust and noise. Live monitoring data were provided via a website to each local authority, enabling users to assess the impact of construction on the local community at any one point or to examine trends over time.

In particular, the partnership with the Environment Agency (EA) is a model of 'modern regulation' that delivered significant benefits to both parties. A dedicated T5 project

manager was seconded to the project from the EA, managing a team of 12 functional specialists covering, for example, waste, contaminated land, water quality, flood defence, river engineering, ecology and consents.

Throughout the project, the EA's approach was to focus on the environmental risks involved and to work with BAA and its construction partners to ensure that the most effective regulation with the minimum regulatory burden was applied. The emphasis was on protecting and enhancing the environment, based on risks and opportunities, by being innovative and following best practice from elsewhere in the construction industry and other sectors. This model is currently being applied to other major projects elsewhere in the UK.

Conclusion

Heathrow T5 has provided the opportunity for BAA and its project partners to set and deliver new standards in environmental sustainability for the construction industry. Lessons learnt have already been incorporated into both complex and small-scale projects within BAA and elsewhere. The challenge was to design, construct and operate T5 in a sustainable manner alongside programme, cost, health and safety and quality requirements. Strong leadership, a challenging brief, effective risk management and pioneering partnerships with regulators and suppliers helped to deliver this.

The T5 journey has been interesting, with a plethora of challenges, opportunities, solutions and learning. Project teams have demonstrated maximum commitment to deliver their brief and enhance the environmental performance of T5.

Acknowledgements

The author thanks BAA for giving permission to publish this paper, Nicholas Bailey of Hyland Edgar Driver for his input into the landscape case study, Boris David of Black & Veatch for his input into the combined water solution case study and the T5 environment team and champions for their remarkable efforts over the years.

What do you think?

If you would like to comment on this paper, please email up to 200 words to the editor at journals@ice.org.uk.

If you would like to write a paper of 2000 to 3500 words about your own experience in this or any related area of civil engineering, the editor will be happy to provide any help or advice you need.

Proceedings of ICE
Civil Engineering 161 May 2008
Pages 25–29 Paper 700043

doi: 10.1680/cien.2007.161.5.25

Keywords
environment; river engineering;
waterways & canals

David Palmer
BEng, CEng, MICE

is an associate at Buro Happold

Phil Wilbraham
BSc, CEng, MICE, MIHT

is head of build, T5, BAA

Heathrow Terminal 5: twin rivers diversion

Moving two rivers from the middle of the Terminal 5 project site at London's Heathrow airport to a new alignment around the western airport perimeter was a critical and highly environmentally sensitive sub-project of the T5 programme. The rivers diversion had to be completed before the original river structures could be demolished to enable continuation of the main terminal development. As described in this paper, the £45 million diversion project included the creation of two 3 km long river channels, phased realignment of 3 km of highway and landscape works to the western boundary of Heathrow.

The Duke of Northumberland's River and Longford River, both of which flowed through the middle of the Terminal 5 (T5) project site at London's Heathrow airport, have long and illustrious histories. Both were constructed by royal command for the purpose of supplying water to royal estates in west London.

It is not known exactly when the Duke of Northumberland's River was first cut, but the current upstream alignment off the River Colne is known to have been commissioned during the reign of Henry VIII (1509–1547). The river was owned for many years by various Dukes of Northumberland, whose family bought it from King James I around 1605. The river was once known as the Old River to distinguish it from the nearby New River or King's River, now known as the Longford River.

Longford River was commissioned by Charles I who, in 1638, commissioned an inquiry into 'how the waters of the River Colne could be brought over Hounslow Heath into the Park' so as to improve the water supply of Hampton Court Palace.

The Duke of Northumberland's River remained in the ownership of the Duke until the 1930s; after a series of changes, it passed to the Environment Agency in 1996. The Longford River is still owned by the Crown and managed by the Royal Parks Agency.

Both rivers diverged from the River Colne to the north of Heathrow airport, with flows in open channels until they reached the northern perimeter of the airfield. In 1947 the flows were conveyed beneath the new northern and southern runways in two pairs of 1·8 m diameter concrete inverted siphons, approximately 500 m long. Between the runways, the twin rivers were conveyed in separate trapezoidal channels for 1 km, with the Longford River being concrete lined through the former Perry Oaks sewage sludge treatment works managed by Thames Water (Fig. 1(a)).

Downstream of the southern runway, both rivers continue in open channels. The Duke of Northumberland's River converges with the east-flowing River Crane for around 5 km, then diverts north through Mogden sewage works to Isleworth Mill and Syon House on the Thames. Longford River flows south east to Bushy Park and Hampton Court Palace, also on the Thames.

Planning process

The issue of the twin rivers was one of the key reasons why the planning process for T5 was so protracted. At the time of the public inquiry, it was proposed both rivers would be diverted around the western perimeter of the airport in a single channel. While this solution would be capable of providing adequate capacity for peak flows, it was not able to deliver low flows to each river downstream without the use of pumps to separate flows. An alternative was therefore required.

Of the options identified, perhaps the most obvious, but also the most controversial, was to

Fig. 1. (a) The two old man-made rivers were ducted under Heathrow airport's main runways in 1947 via 500 m long inverted siphons (shown red); these were linked by open channels right through the middle of the T5 site. (b) The diverted rivers run alongside the realigned western perimeter road, with a culvert under the welcome roundabout (marked T5) and inverted siphon under the southwest airport roundabout (bottom left)

place both rivers in tunnels (inverted siphons) under the proposed main terminal building. Although this meant that the rivers would provide almost no ecological and environmental value, it was not without precedent as the rivers were already in siphons beneath the north and south runways. The scheme was developed to pre-detailed design stage and various studies were undertaken to asses what degree of environmental value, if any, could be provided—these studies included research into whether fish would migrate in siphons of lengths up to 2 km.

In early 2001, owner BAA proposed that the tunnel scheme be put forward. The Environment Agency confirmed it was disappointed with this approach and would oppose any variation to the planning application and possibly even reopen the planning inquiry on the grounds that the proposal represented a significant departure from the original scheme.

The 'clash' of ideals as to how the rivers could best be treated represented a very significant risk to BAA in terms of further extending an already protracted planning process. It was thus decided that the most efficient way forward would be to identify a new scheme that met the key criteria not just of BAA but also the key stakeholders—the Environment Agency, Royal Parks Agency, Civil Aviation Authority (CAA), National Air Traffic Services and the planning authorities of London Borough of Hillingdon and Spelthorne Borough Council.

A series of meetings was held with all the key parties involved. A scheme was identified that involved diverting the twin rivers in open channels around the western perimeter of the airport. The scheme, described in detail in this paper and referred to as the 'twin rivers diversion', was capable of meeting hydraulic criteria while delivering significant environmental enhancement when measured against the existing rivers (Fig. 1(b)). Although the twin rivers diversion meant the loss of some valuable airfield, it did have the approval in principle of the key parties and allowed a planning application for T5 to be submitted to the Government Office of London in August 2001, which included the twin rivers diversion scheme in outline.

The Government Office of London granted a 'Grampian' permission for T5 in October 2001, which meant that BAA could begin construction of T5 once the twin rivers diversion had been formally approved as a separate, detailed planning application. The twin rivers diversion planning application was submitted to Hounslow council in January 2002 and planning approval was granted in July 2002.

Design

All design work was carried out through close working with the key stakeholders, the Environment Agency and the Royal Parks Agency, in an informal but nonetheless very effective partnered relationship. Monthly meetings and the relationships that were then forged were integral to the success of the project; the commitment and enthusiasm of the Environment Agency and the Royal Parks Agency were important factors in this.

The twin rivers diversion design team comprised consultants from Black & Veatch, KBR, Hyland Edgar Driver and TPS Consult supporting the BAA team, who were formed into sub-teams to cover the rivers, structures, roads, services and landscape. Later in the design process, the team was supplemented by Lang

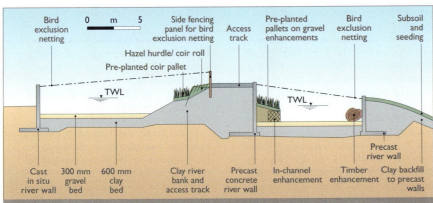

Fig. 2. Typical section through the twin rivers diversion—each river is typically defined by one vertical concrete wall and a two-stage naturalised bank

Fig. 3. The central 4 m wide access track was built on a T-bar either side of the welcome roundabout culverts to allow more lateral space for landscaped reinforced-earth banks. The bird exclusion mesh, which had to be installed along the entire length, can also be seen

O'Rourke professionals, who provided support in the form of design reviews, costings, phasing and buildability advice.

In terms of the rivers, the primary criteria were twofold

■ to ensure that the Duke of Northumberland's River and Longford River had sufficient capacity to convey peak flows of 3 m³/s and 1·5 m³/s respectively
■ to ensure that the diverted rivers provided a minimum of environmental equivalence when compared with the rivers on the existing pre-T5 alignment.

Designing twin canals that convey relatively low peak flows is straightforward. The main challenge lay in ensuring that the diverted rivers provided significant ecological and environmental enhancement without compromising the narrow corridor that was the western perimeter.

Both rivers are diverted in open channels running parallel to each other with a 4 m wide access track for maintenance purposes between the two watercourses. Where space is particularly constrained at the western end of the north and south runways, the rivers are defined by vertical concrete walls. Elsewhere, the rivers are defined by one vertical concrete wall and a two-stage naturalised bank (Fig. 2).

The rivers were generally kept in open channel to maximise their environmental value. The only exceptions were at the welcome roundabout and the south west airport roundabout, where they were respectively conveyed underground in a 120 m long, free-surface, precast culvert and a twin 65 m long, 2·4 m diameter, precast concrete pipe inverted siphon.

An opportunity arose to provide a more environmentally landscaped approach either side of the welcome roundabout using reinforced-earth banks in place of concrete walls. Additional space for the slopes required reducing the cross-section of the diverted rivers, and this had to be done without compromising the hydraulic

capacity. The solution was to form the central access track on a reinforced concrete T-bar (Fig. 3).

At the upstream end of each diverted river, a concrete-lined sediment basin was designed to reduce normal velocities by approximately 80%. The rivers are defined by very shallow gradients, low velocities and high sediment loads, so the basins were designed to minimise maintenance issues related to sedimentation and loss of conveyance within the channels in years to come.

A key part of the design related to the formation of the river walls. Due to the land form of the western side of the airport, levels in the diversion are perched above the realigned western perimeter road. Of the total length of over 12 km of river bank, approximately 9 km is thus formed by vertical walls. For T5, there was also a project-wide objective to construct using units manufactured off-site as far as possible—meeting this aspiration was made more challenging by the need for watertight joints.

The right bank of the Duke of Northumber-

land's River and both banks of the Longford River could be backfilled with as-dug clay and so the risks associated with precast walls were much lower. The key area of consideration was the left bank of the Duke of Northumberland's River, with water levels higher than the western perimeter road and no room to backfill the river with clay. After much consideration it was decided that these walls would be formed using traditional in-situ methods with heavy-duty water-bars at all construction joints.

In all, a total of 3700 m of in-situ wall was cast on site, with over 5 km of river wall being formed by 10 m long, 20 t, precast sections. The precast walls were joined together with a 600 mm in-situ stitch incorporating a water-bar and hydrophilic strip (Fig. 4).

Environmental enhancements

Having designed the channels such that 95% of their length was in open channel, a suite of environmental enhancements and ecological

Fig. 4. Over half of the 9 km of concrete river walls comprises 10 m long, 20 t precast units joined together with a 600 mm in-situ stitch incorporating a water-bar and hydrophilic strip

features was identified and included to ensure the diverted rivers exceeded the original rivers in terms of environmental value.

The naturalised banks of the diverted rivers were formed using London Clay won from the major excavation works on the terminal site. The river bed was formed by a 300 mm deep as-dug terrace gravel, again won from the terminal site excavations. This meant that, while it was accepted that the twin rivers diversion was providing canals rather than natural rivers, they were being formed using the same geology that characterised other rivers in the Thames and Colne Valley.

The banks were two-stage, with a 1 m berm at typical water level and an upper slope planted with a wildflower seed mix. At the rear of the berms, coir rolls and hazel hurdles created shallow vertical shelves to provide habitats for small mammals such as water voles.

There were three principal types of in-channel enhancement

- gabions filled with terrace gravel providing infrastructure for the pre-planted pallets
- gabions filled with rocks providing habitat for macro-invertebrates and small fish
- timber logs, weighted to the bed, providing habitat for fish and invertebrates.

The enhancements were positioned within the channels, alternating on opposite banks so as to create a degree of sinuosity within the rivers and local velocity variations at pinch

points. These measures further enhanced biodiversity (Fig. 5).

The logs, mainly mature willow trunks, were all donated by the Royal Parks Agency and British Airways following tree felling programmes at parks in west London and Waterside to the north of the airport respectively—another example of successful partnering.

Planting within the twin rivers diversion was largely carried out using pre-planted pallets. These are coir mats, typically 100 mm deep and provided in 1 m × 2 m plates. By planting in this way, plants could be established a full 18 months before installation. This greatly speeded up the planting process on site and, most importantly, minimised the loss of plant stock as they were reasonably mature at the time of installation. While the capital costs of planting in this way may be marginally higher, whole-life costs demonstrate this was by far the most efficient way of planting. In all, over 8000 m² of pre-planted pallets were installed, consisting of over 84 000 plants in three broad groupings of aquatic, emergent and marshland-type species. Over 36 different species were planted, all of which were native to the Colne Valley.

A massive commitment to landscape was made within the western corridor of the airport. As well as supplementing the rivers and roads, the landscape provides a transition between Colne Valley to the west of the airport and T5. This was achieved by planting over 450 semi-mature trees, 2000 semi-mature native shrubs and 100 000 evergreen groundcover shrubs.[1]

Bird exclusion measures

The creation of two new rivers adjacent to the world's busiest airport created challenges unique to this environment. Open bodies of water such as the twin rivers diversion are a significant attraction to birds—this, in turn, represents a significant hazard to aircraft safety in the form of mid-air bird-strikes.

One of the key conditions of the planning agreement was that the twin rivers diversion must include a bird exclusion system, as yet undefined, to the satisfaction of the CAA. The CAA's brief for this was quite clear: exclude all hazardous birds from the twin rivers diversion. This was complicated slightly by a different condition in the planning agreement, this time inserted at the behest of the Environment Agency and Royal Parks Agency: the exclusion must represent a minimal impact in terms of ecology. In addition, any system had to allow easy and safe maintenance of the rivers.

A number of potential systems were identified and considered, ranging from a retractable steel mesh canopy to total coverage with floating plastic balls, as used on other ponds at the airport. Bird balls were not appropriate in this case as they would exclude light from the rivers and preclude all plant and most other life forms.

After a number of field tests, a 0·3 mm diameter, lightweight polypropylene netting with a mesh size of 75 mm was identified as the best solution as it excluded all hazardous birds and yet allowed exit and entry to a range of invertebrates up to, and including, the emperor

Fig. 5. Diverted river beds were made from locally excavated terrace gravel, which was also used in gabions to create alternating berms along with rock-filled gabions and logs

Fig. 6. View of the completed twin rivers diversion and realigned western perimeter road looking north towards the new welcome roundabout

dragonfly. The lightweight mesh was also relatively inexpensive, very easy to install and replace, and flimsy enough in appearance so as not to attract trespassers to use it as means of crossing the rivers (Fig. 3).

Construction process

The new 3 km rivers were constructed over the old western perimeter road, a busy airport road that therefore required realigning eastwards towards the airport (Fig. 6). The road was realigned in six permanent and two temporary phases; it was kept open to traffic throughout. The project included the construction of a 130 m long arch bridge under the new welcome roundabout.

The construction team peaked at 400 members, working closely to ensure a safe environment in a very constrained site between two live roads. Other restrictions in this unique working environment included limitations placed on the use of cranes due to landing aircraft. Construction was carried out over an 18-month period consisting of two winters and one summer.

Switching flows and ecology

Once construction of the new off-line channels was complete, the next step was to fill them with water. This was done by abstracting from the adjacent existing river at a rate no greater than 10% of the flow in the river. The new channels were filled to peak design water level and this was maintained for up to a week to ensure that there were no leaks. Water levels were then lowered to the same level as the downstream point at which they would connect into the existing river. The process of switching flows was to

- remove bund separating the new channel from the old river at the downstream end of the new channel (old river continues to flow)
- remove bund separating the new channel from the old river at the upstream end (both new channel and old rivers now flowing)
- install new bunds across head and tail of the old river (only new channel flowing).

The switch-over days were among the most exciting on site for the project team and represented the culmination of many years of hard work. Flows in the Longford River and Duke of Northumberland's River were switched in February and April 2004 respectively, and both operations went ahead without incident.

Decommissioning the old rivers could then begin. The first process was to carry out a series of fish captures along the now off-line old rivers. The fish were caught using the standard

electro-fishing method and released into the River Colne, between the head of the Longford River and the Duke of Northumberland's River, at Waterside park owned by British Airways. A total of three passes were carried out in the Longford River and four in the Duke of Northumberland's River in April 2004, resulting in the rescue of more than 40 000 fish.

Following diversion of the Duke of Northumberland's River, before the old river was dewatered, the top 150 mm of river bed material was excavated and used to 'seed' both of the new diverted rivers in 0·5 m³ deposits every 50–75 m throughout the courses of the new channels. This helped to 'kick-start' the ecology of the new channels as the gravels of the old rivers were rich in macro-invertebrates (such as freshwater shrimp, water hog-louse, damselfly nymphs, leaches and snails) and freshwater mussels. Swan and duck variety mussels were hand-picked from the bed of the old Duke of Northumberland's River and a total of over 1200 translocated to the new channels.

The process of translocation from the original rivers had actually begun several years before the rivers were switched. In 2000, it was established that a sizeable colony of water voles was present in the twin rivers through the former Thames Water sewage sludge works. The voles were removed from the twin rivers in autumn 2001, bred in captivity and released into the wild along the Stanwell Moor Ditch at Staines Moor site of special scientific interest, where they have continued to thrive.

The process of translocating plants from the old twin rivers began in October 2002 following a survey of flora along the rivers. The species native to Colne Valley were lesser pond-sedge (*Carex acutiformus*) and great yellow cress (*Rorippa amphibia*), and samples of each were used within the pre-planted pallets along with 35 other native plant species.

The river switch process required four land-drainage consents, two abstraction licences and two discharge consents, all required by the Environment Agency, as well as ownership consent from the Environment Agency and Royal Parks Agency.

The old twin river courses were finally handed over to the T5 earthworks team three weeks ahead of programme and £5 million under budget (see Table 1). Following initial deconstruction, archaeological searches were carried out and then all traces of the rivers demolished.

Safeguarding the twin rivers diversion

The story of the twin rivers did not end in 2004. One condition of the 2002 planning permission was that a body should be formed to ensure continued management of the rivers. The Twin Rivers Management Committee was formed and comprises equal representation of

BAA, the Environment Agency and Royal Parks Agency. The committee meets twice every year and, among other things, oversees a twice-yearly ecological survey of the twin rivers diversion to monitor its performance and identify any remedial works required.

Summary

The diversion of the twin rivers was a hugely successful project within the T5 programme. Crucial to this success was the way in which BAA and its suppliers formed a highly motivated and dedicated design and construction team, which was co-located at Heathrow.

The team worked in partnership with key stakeholders to ensure that an integrated river, road and landscape corridor was delivered that not only enhances the ecological value of the rivers and the western perimeter of the airport but, by keeping the western perimeter road open throughout, also ensured no significant impact on Heathrow airport or local communities.

The twin rivers diversion achieved a number of awards, including a 92% 'excellent' rating in the UK's civil engineering environmental quality assessment award scheme, and shortlisting for the 2004 British construction industry awards.

Table 1. Twin rivers diversion costs		Cost: £ million
Miscellaneous	Staff costs, insurance, etc.	14·7
River works	Walls	12·5
	Structures	1·4
	Sediment basins	1·8
	Enhancements	0·7
	River planting and soft engineering	0·3
	Bird exclusion works	0·1
	Rivers switch	0·15
Roads	Permanent and temporary diversions	8·4
Services	Permanent and temporary diversions	1
Landscape	Soft and hard landscaping	4·3
Total		£45·35

References

1. LISTER B. Heathrow Terminal 5: enhancing environmental sustainability. *Proceedings of the Institution of Civil Engineers, Civil Engineering*, 2008, **161**, Special Issue—Heathrow Airport Terminal 5, May, 21–24.

What do you think?

If you would like to comment on this paper, please email up to 200 words to the editor at journals@ice.org.uk.

If you would like to write a paper of 2000 to 3500 words about your own experience in this or any related area of civil engineering, the editor will be happy to provide any help or advice you need.

Proceedings of ICE
Civil Engineering 161 May 2008
Pages 30–37 Paper 700060

doi: 10.1680/cien.2007.161.5.30

Keywords
airports; geotechnical engineering;
tunnels & tunnelling

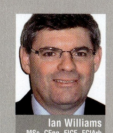

Ian Williams
MSc, CEng, FICE, FCIArb

is programme delivery
manager at BAA

Heathrow Terminal 5: tunnelled underground infrastructure

Terminal 5 at London's Heathrow airport was one of the largest infrastructure projects in Europe over the last five years. It included approximately 14 km of tunnelling to provide underground infrastructure for rail, road and effluent discharge. In all cases the tunnels were delivered without major incident, thanks to a single integrated team which made all aspects of risk management central to its delivery philosophy. This paper describes the four key projects for which the key aspect of delivery was tunnelling – the airside road, the storm water outfall and the Piccadilly line and Heathrow Express extensions. It discusses the range of aspects that made up the delivery of this infrastructure, ranging from design through to performance of tunnelling machines and the control of surface settlement.

A key infrastructure element of the £4·3 billion Terminal 5 (T5) development at London's Heathrow airport is approximately 14 km of tunnelling to provide the terminal with road, underground rail links and drainage provision. The tunnels range in length from 1·3 km to 4·1 km and from 3 m to 8·1 m in diameter. The majority were constructed using tunnel boring machines (TBMs) and innovative developments of a single-shell, shotcrete-lined method called Lasershell[1] for construction of the numerous short, complex underground structures such as shaft and cross-passage complexes.

Ground conditions

The ground conditions at Heathrow are consistent over the area and consist of made ground underlain by Terrace Gravels over London Clay. The gravel–clay interface lies between 3–10 m below the ground surface, and the ground water level lies in the gravel about 2 m below the surface.

To take advantage of the benefits of the London Clay as a tunnelling medium, all the tunnels were designed so that the tunnels would be excavated through clay.

Tunnel projects

The plan location of the various T5 tunnel projects is shown on Fig. 1[2] and Table 1 provides a summary of the main tunnel lengths and diameters.

Airside road tunnel
The purpose of the airside road tunnel is to provide a fixed, all-weather underground road link between T5 and the remote aircraft stands and the existing central terminal area (Terminals

Table 1. Summary of tunnel lengths and diameters			
Tunnel project	**Length: m**	**Bores**	**Inside diameter: m**
Airside road tunnel	1270	2	8·1
Storm water outfall tunnel	4100	1	3·0
Piccadilly line extension	1600	2	4·5
Heathrow Express extension	1700	2	5·675

Institution of Civil Engineers

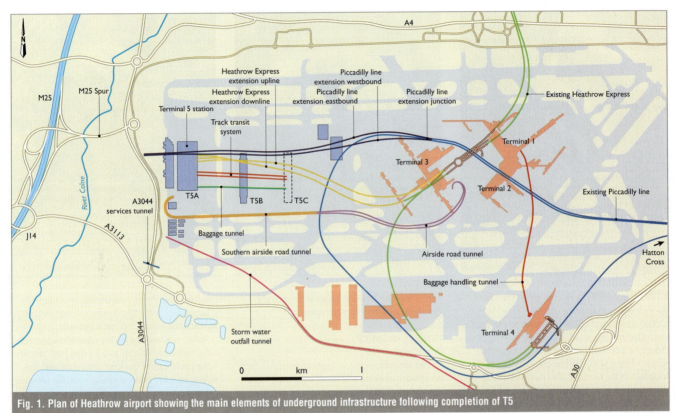

Fig. 1. Plan of Heathrow airport showing the main elements of underground infrastructure following completion of T5

1, 2 and 3). The project consisted of 1·3 km twin-bored tunnels with an internal diameter of 8·1 m (Fig. 2). At each end, cut-and-cover portal structures were constructed using the observational method.[3,4] Twelve emergency cross-passages were constructed to connect each bore, with an approximate linear spacing of 130 m. In addition, two mid-tunnel drainage sumps and two underground mechanical and electrical equipment rooms were also excavated.

A balance had to be struck between the requirements of the user, Heathrow Airport Limited, and the buildability of the scheme. To meet the operational requirements the position of both portals was fixed, and it was chosen to have a 6 m wide carriageway with one-directional travel. This resulted in the requirement for two 8·1 m internal diameter parallel tunnels. Vehicles intended to use the tunnel could only operate at gradients up to 1 in 19, which fixed the vertical alignment constraints for the tunnel. An additional operational requirement was that the tunnel needed to be as short as possible to reduce journey time.

There were also a number of constraints on route selection, probably the most significant being that the tunnel would pass beneath critical airport infrastructure, including aircraft piers and strategic airport fuel and fire mains. The main construction challenge therefore would be maintaining ground movements caused by the tunnelling to a level that would not affect the

overlying infrastructure.[5]

A further constraint was that the tunnel would have to pass beneath or over the existing Heathrow Express railway tunnel. The design gradient and the potential increased depth ruled out going beneath it, so the resulting horizontal alignment has a flat 'W' profile with the portals and the Heathrow Express being high points in the tunnel with two low points either side of the crossing.

To minimise settlement, the ideal tunnelling horizon was the London Clay (stiff over consolidated clay), but at the portals there was a high potential that the crown of the tunnel might encroach into the Terrace Gravel. Analysis of the structure and infrastructure indicated that settlement needed to be maintained below 25 mm to ensure the assets remained within safe limits.

Storm water outfall tunnel

The purpose of the storm water outfall tunnel is to provide a conduit in which the surface run-off from the T5 structures and the paved areas (stands and taxiways) could be discharged into the Clockhouse Lane Pit, around 2 km south of T5.[6] The tunnel was approximately 4·1 km and had an internal diameter of 2·91 m to accommodate the flow requirements.

In addition to the tunnel, inlet and outlet shafts of 10·5 m diameter and two intermediate shafts of 6 m diameter were constructed – the latter being for operational maintenance

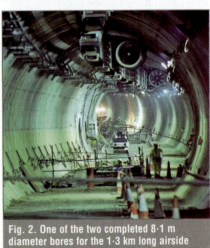

Fig. 2. One of the two completed 8·1 m diameter bores for the 1·3 km long airside road tunnel

intervention.

The key ground movement challenge was passing beneath the end of the southern runway.

Piccadilly line extension

A fundamental condition on which the secretary of state gave approval for T5 was that a rail link must be in place to serve T5. The existing Piccadilly line infrastructure to Heathrow airport consisted of a loop line with underground stations at the central terminal area and Terminal 4. The extension of the Piccadilly line

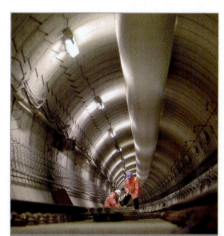

Fig. 3. One of the two completed 4·5 m diameter bores for the 1·6 km long Piccadilly line extension tunnel

Fig. 4. New sprayed-concrete-lined tunnels connect both the existing and new Piccadilly line tunnels to the new 23 m wide diaphragm-walled junction box

Fig. 5. One of the two completed 5·5 m diameter bores for the 1·7 km long Heathrow Express tunnel

to T5 consisted of approximately 1·6 km twin tunnels of 4·5 m internal diameter (Fig. 3).

The connection of the extension to the existing infrastructure was formed 300 m west of the existing central terminal area at the step-plate junction formed during the construction of the loop line from Terminal 4 in the 1980s. The connection between the existing and new tunnels consisted of a 17 m deep, 40 m long and 23 m wide sub-structure constructed using diaphragm-wall techniques and four short sprayed-concrete-lined connection tunnels (Fig. 4).

The Piccadilly line extension tunnels were driven from a cut-and-cover launch chamber approximately 300 m west of the first satellite terminal, T5B. The bored tunnels were connected to the station structure at T5's main terminal building T5A by cut-and-cover structures.

The main challenges to this project were controlling ground movement of surface structures such as the Heathrow fuel farm and taxiway infrastructure. In addition, one of the bored tunnels passes beneath the existing Piccadilly loop line tunnel and both tunnels were driven parallel to the existing step-plate tunnels at the central terminal area.

Heathrow Express extension

The existing Heathrow Express line linked the central terminal area to Terminal 4. During the construction of this line in the mid-1990s, stub tunnels off the tunnels to Terminal 4 were constructed approximately 200 m west of the central terminal area station as safeguard provision to extend the line towards the proposed T5 site.

The extension consists of approximately 1·7 km twin tunnels of 5·5 m diameter (Fig. 5). They were driven from an adjacent cut-and-cover launch chamber to the Piccadilly line extension tunnels. Similar to the Piccadilly line extension tunnels, the challenge was controlling ground and structure movement, and in this

case it was the crossing beneath the existing Piccadilly loop line by both tunnels.

Project delivery

The tunnelling project team consisted of five companies working within an integrated team framework. The companies included BAA (client and project leadership), Morgan Vinci JV (tunnel constructor), Mott MacDonald (tunnel and civil designer), Beton-und Montierbau (sprayed-concrete lining designer and site supervision) and Laing O'Rourke (civil works for the airside road tunnel portals).

Risk management

From the experience of other projects[7] and BAA's own experience of the Heathrow Express tunnel collapse in 1994,[8] managing risk was made central to the management process adopted. The strategy for managing risk was through integrating risk and delivery management.

Initially the process involved exposing the risks through project-risk reviews and thereafter regularly updating and reviewing 'live' risk registers and live risk-response plans within the framework of a project-execution plan.

Each of the specific tunnel projects presented a number of challenges, the two most significant physical challenges being controlling ground movement (and eliminating tunnel collapse) and workforce safety. It was considered that if these two areas were managed successfully, the other key risks of time and cost would directly be influenced positively.

The team adopted the fundamental principle of providing control measures rather than consequence control measures as a basic risk-management approach to achieve the optimum solution. Specifically, the team concentrated on eliminating the risk – for example ground movement and face collapse – by implementing control

measures and thereby preventing or minimising it happening in the first instance rather than relying on a consequence-control measure.

While not mandatory, the team implemented – through commitment to adopting best practice – the guidelines of the joint insurance and tunnelling industry code of practice for the procurement, design and construction of tunnels and associated underground structures in the UK.[9]

Assurance scheme

All aspects of the tunnels were delivered under each of the company's self-certification system. In addition, further levels of assurance were put in place: first on the safety of the works and second with respect to the process and product (design and construction).

The objective of the works safety assurance was to eliminate the occurrence of a major incident, such as a major ground failure. This initiative built on BAA's philosophy of risk management being central to the success of the project. The assurance was provided by a small number of individuals from Mott MacDonald and Beton-und Montierbau, whose responsibility was to review proposals and monitor the execution of the construction works. While these people were integrated into the delivery team, their line accountability was straight through to BAA and they did not have any responsibility for cost or programme aspects.

The project team developed an assurance system for the Piccadilly line extension with respect to the process and product to meet owner London Underground's requirements.[10] This system was also extended to cover all tunnel works. The process covered design and construction and provided the project team with a subsystem that demonstrated compliance throughout the entire process. It also provided the 'backbone' of information for the quality records for the health and safety file, and was

Fig. 6. The airside road tunnel was bored using a purpose-designed-and-built, dual-mode TBM by Herrenknecht

Fig. 7. A refurbished closed-face Lovat TBM was used for the storm water outfall tunnel

Fig. 8. The Piccadilly line extension was bored using refurbished open-faced TBM originally manufactured by Dosco UK

used to good effect when submissions were being made in seeking permission to enter into the zone of influence of third-party structures.

Safety management

A central project objective was ensuring workforce safety. A number of innovations and initiatives were developed and followed on T5 to meet this objective. Adoption of cause-control measures and targeting attitudes to personal safety were the two key elements of the strategy. It has long been considered that significant improvement in safety can only be achieved if a positive safety culture exists in everyone. To address this, a behavioural safety initiative was introduced: the 'incident and injury free' programme.[11]

The key element of the programme was to make safety a personal commitment, moving safety to a core value and creating a vision and expectation that all incidents and injuries are avoidable. The team investigated and developed methods and techniques to reduce workforce exposure.

One of the main innovations involved further developments with the sprayed-concrete-lining-technique. The Lasershell method was developed to improve safety by eliminating the need

for miners to enter an unsupported face area and to mechanise the process to reduce injuries.

Construction methods

The majority of the tunnels were constructed using a TBM. The sprayed-concrete-lining method was used to construct the numerous short lengths of geometrically complex structures such as shaft and tunnel connections and cross-passages.

TBMs

The TBM type was governed by suitability to excavate London Clay below the water-bearing gravel interface. Four TBMs were used, consisting of two closed-faced and two open-faced TBMs.

The airside road tunnel was bored using a purpose-designed-and-built, dual-mode TBM by Herrenknecht (Fig. 6)[5] and the storm water outfall tunnel was bored using a refurbished closed-face Lovat machine (Fig. 7).

The Piccadilly line and Heathrow Express extensions were both bored using refurbished open-faced TBMs originally manufactured by Dosco (UK) (Figs 8 and 9). The excavation mode for these was a backhoe and boom-

mounted roadheader, and both were equipped with face-breasting plates to provide limited face-stability control.

Details of all TBMs are given in Table 2. The refurbished machines did not present any procurement challenge whereas the airside road tunnel TBM required considerable design development[5] and manufacturing time, which was programmed into the overall schedule. Extending the single integrated team to Herrenknecht was the key to the success of this aspect of the project.

On completion of the projects, the TBMs were returned to the original supplier or manufacturer.

TBM spoil removal method and segment delivery

Spoil removal from the TBMs was done by tunnel conveyor, except in the storm water outfall tunnel where it was removed via muck skips to the shaft bottom. Removal of spoil from

Fig. 9. Refurbished open-faced TBM originally manufactured by Dosco being removed after completing the Heathrow Express extension

Characteristic	Airside road tunnel	Storm water outfall tunnel	Piccadilly line extension	Heathrow Express extension
TBM type	Dual mode EPBM	Drum digger	Cutter boom (roadheader)	Backhoe mounted on boom
TBM manufacturer	Herrenknecht	Lovat	Dosco	Dosco
External diameter: mm	9160	3345	4810	6125
Overall TBM length: mm	10360	4035	4570	–
Length/diameter ratio	1·13	1·20	0·95	–
Overall length (including back-up): m	–	44·2	56·5	42·8
Cutterhead power: kW	2800	373	142 (roadheader cutter)	N/A
Cutterhead motors	7	6	1	N/A
Cutterhead rotation speed: rpm	0–3	–	27 or 50	N/A
Rams: no.	28	12	16	–
Total capacity: kN	69272	1020	–	–
Stroke: mm	2500	1680	1680	–

Table 2. Summary of TBM characteristics

launch chambers was done using high-angle conveyors (Fig. 10), the exception in this case being the airside road tunnel where the tunnel conveyor was extended out of the tunnel portal. Apart from minor mechanical breakdowns, the tunnel and high-angle conveyors performed well and the historical concern of material 'sticking' was not experienced.

A key factor in the selection of segment transportation for the airside road tunnel was the maximum 5·3 % tunnel gradient. The main benefit of the system selected – a rubber-wheeled transporter supplied by Paolo de Nicola (Fig. 11) – was the short stopping distance and the ability for the driver to operate the vehicle from either end, enabling unimpaired vision. The same vehicle was also used to service the Heathrow Express extension drives. Rail vehicles were used for the Piccadilly line extension and storm water outfall tunnel drives.

Sprayed-concrete-lined structures

The Laseshell sprayed-concrete-lining method developed by Morgan Est and Beton-und Montierbau was used for constructing the numerous short-length, complex-geometry tunnels such as cross-passages and shaft intersections (Fig. 12).[12]

The main characteristics of Lasershell are a circular cross-section with an inclined domed face (Fig. 13) and rapid shotcrete ring closure, and the elimination of wire mesh and lattice girders. The elimination of lattice girders required the development of an innovative survey tool to control the tunnel profile and shotcrete thickness.[12]

The method involves a permanent shotcrete lining applied in three layers. The initial layer, approximately 75 mm thick, is applied for face sealing immediately. The primary layer – the main structure element – is applied in a single cycle to the required design thickness. Both the initial and primary layers contain steel and polypropylene fibres. On completion of the excavation works, a nominal 50 mm thick trowel-finish layer without fibres is applied.

The excavation and spoil-removal plant was selected for the projects primarily based on size of tunnel to be excavated. The key excavation plant was a specific mining excavator supplied by Schaeff.

An extensive testing programme was carried out to demonstrate the long-term durability and structural integrity of the shotcrete to achieve the design life of 120 years. In addition to the standard strength testing, the programme included tests to determine permeability, dynamic modulus and petrographic composition

Fig. 10. Spoil removal for the Piccadilly line extension was undertaken by a conveyor in the tunnel and a high-angle conveyor up the side of the launch chamber shaft

Fig. 11. Rubber-wheeled transporter used for segment delivery in the airside road and Heathrow Express extension tunnels – the vehicle could be operated from either end

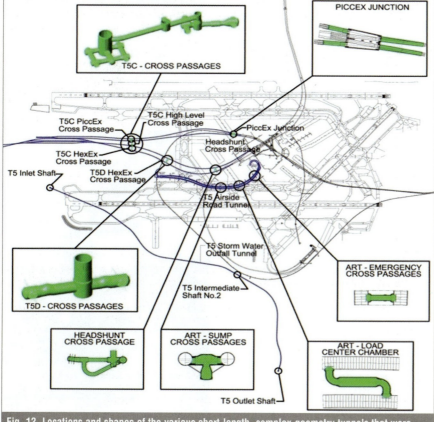

Fig. 12. Locations and shapes of the various short-length, complex-geometry tunnels that were built using the Lasershell sprayed-concrete-lining method

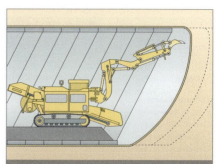

Fig. 13. The Lasershell method involves excavating short lengths of a circular tunnel with an inclined, domed face and then immediately sealing the excavation with 75 mm of shotcrete, followed by a single-pass structural layer and 50 mm finish layer

best practice and targets and define BAA policy. These strategies challenged BAA's design standards and embedded sustainability into future capital developments.

As the scheme design evolved, performance against these requirements was continuously reviewed and recommendations for improvements made. Towards the end of the design stage, all sustainability documentation was passed on to the project teams to take forward into construction.

Construction stage

During production design and prior to starting on site, an environmental review of all sub-projects captured any previously outlined sustainability requirements together with newly identified risks and opportunities associated with the construction stage, such as

- ecological constraints
- permit, licensing or consent requirements

- location sensitivities
- night work requirements
- activity-related impacts.

Clear responsibility for impact mitigation was achieved through individual project team environment plans, which became crucial live documents in the management of site activities. Throughout the terminal construction, regular environmental reviews, audits and inspections and monthly performance reporting helped to provide assurance that activities were being properly managed and the original brief fulfilled.

As commissioning progressed, residual environmental requirements (e.g. discharge consents and energy targets) were handed over to BAA's operational teams to ensure that any regulatory, policy or operational requirements could be achieved throughout the lifetime of T5, and ultimately into its decommissioning.

The performance of T5 against the project sustainability criteria may best be illustrated through a series of case studies.

Combined water solution

Delivering the T5 environmental target of '70% reduction in potable water demand' required an initial understanding of the different water quality and quantity needs within the terminal facility. This resulted in a 'fit for use' water solution, combining separate potable and non-potable water supply systems.

Potable water sourced from the local water provider is used for activities such as catering and showers, complemented by a non-potable source for activities such as toilet flushing, vehicle washing and fire fighting on aircraft stands. The non-potable element is supplied from three sources.

- *Groundwater abstraction.* A total of 60% of T5's non-potable water requirement of over 0·7 million m³ a year is sourced from groundwater abstraction from the Chalk aquifer below T5. Two 150 m deep boreholes (Fig. 2) were drilled in 2003, yielding a combined flow of up to 36 l/s. Borehole water was used on site throughout the construction phase in the supply of site compounds (through a reverse-osmosis filtering system), concrete batchers and dust suppressors, thus minimising the impact of construction on water resources.
- *Rainwater harvesting.* Rainfall runoff from the T5 development totals around 0·9 million m³ per year. It discharges to a local reservoir, Clockhouse Lane Pit (Fig. 3), through a series of oil interceptors, silt traps and continuous quality monitoring for biological oxygen demand and pH. An abstraction licence has been granted to return a maximum of 0·25 million m³ a year to T5, which will account for approximately 35% of its non-potable water requirement.
- *Recycled water.* The final 5% of T5's non-potable water requirement is derived from reverse-osmosis treatment of water used in the air-conditioning system towers. The by-product of this process—water with high concentrations of salts—would normally be discharged to the foul drainage system, but on-site treatment at T5 enables its re-use.

The three sources require minimal treatment, involving filtering and chlorination, prior to storage and distribution. When compared with the use of potable sources, the combined solution is extremely sustainable in terms of energy use, resource efficiency and cost. The reduction in potable water consumption minimises the impact of source development as well as the impact of potable water production and distribution, principally associated with chemical resources, waste generation and pumping energy.

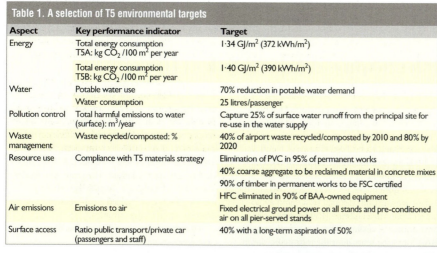

Table 1. A selection of T5 environmental targets

Aspect	Key performance indicator	Target
Energy	Total energy consumption T5A: kg CO_2 /100 m² per year	1·34 GJ/m² (372 kWh/m²)
	Total energy consumption T5B: kg CO_2 /100 m² per year	1·40 GJ/m² (390 kWh/m²)
Water	Potable water use	70% reduction in potable water demand
	Water consumption	25 litres/passenger
Pollution control	Total harmful emissions to water (surface): m³/year	Capture 25% of surface water runoff from the principal site for re-use in the water supply
Waste management	Waste recycled/composted: %	40% of airport waste recycled/composted by 2010 and 80% by 2020
Resource use	Compliance with T5 materials strategy	Elimination of PVC in 95% of permanent works
		40% coarse aggregate to be reclaimed material in concrete mixes
		90% of timber in permanent works to be FSC certified
		HFC eliminated in 90% of BAA-owned equipment
Air emissions	Emissions to air	Fixed electrical ground power on all stands and pre-conditioned air on all pier-served stands
Surface access	Ratio public transport/private car (passengers and staff)	40% with a long-term aspiration of 50%

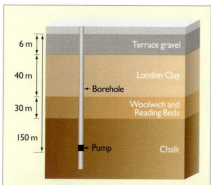

Fig. 2. The primary source of T5's annual 0·7 million m³ non-potable water requirement comes from two 150 m deep boreholes into the Chalk aquifer below the terminal, which yield a combined flow of up to 36 l/s

Fig. 3. All 0·9 million m³ annual rainfall runoff from T5 is ducted via oil interceptors, silt traps and quality monitors to a local reservoir called Clockhouse Lane Pit, from which nearly a third is abstracted back to the terminal; this accounts for 35% of the terminal's non-potable water use

—and involved using one-year old samples.

Table 3 summarises the mix design that met the specification and workability/application requirements. The shotcrete was produced on site at the purpose-built batching plant dedicated to the T5 project.

Recognising the hazards of shotcrete application, most of the shotcrete was applied using robotic spray techniques (Fig. 14). The exception to this was the application of the finish layer using hand-held spray equipment. The Jacon static shotcrete pump and skid-steer-mounted shotcrete robot were used exclusively.

Resources and shift arrangements

The majority of the construction operatives for tunnel construction were directly employed. The exceptions to this were the specialist TBM erection and installation teams.

For the airside road tunnel and storm water outfall tunnel projects, seven-day working was adopted consisting of three shift gangs working on a pattern of six 12 h shifts followed by three days off.

The shift pattern for Piccadilly line and Heathrow Express extensions consisted of ten 12 h shifts (Monday to Friday). Weekend working was dedicated to planned maintenance, apart from when continuous tunnelling was required, such as crossing beneath sensitive existing structures or through permanent shaft locations.

Typical labour resources for each tunnel projects were as follows.

- TBM – ten people per shift
- tunnel, shaft bottom and top operation – seven people per shift
- surface operation – five people per day
- attendance support – 25 people per day.

Design

Pre-cast concrete segments

Most of the precast concrete segments were of an expanded form and fibre-reinforced. The exceptions were the airside road tunnel precast concrete segments and a limited number of bolted segments, which were used for the start of the other tunnel drives and passing through preformed chambers.

The main reason for the airside road tunnel segments being bar-reinforced was the concern about potential cracking under shove loads as a result of the build quality causing point loads on adjacent segments – a phenomenon known as 'out of plane'. A further significant factor was that, at the time of the segment design for airside road tunnel, limited data were available to demonstrate successful utilisation of fibre-reinforced segment technology for similar-sized tunnels.

Table 4 summarises the key characteristics of the segments used.

Piccadilly line extension junction box

The Piccadilly line extension junction box was a significant structure in its own right. The railway design dictated that the connection was to be coincident with the existing step-plate junction, in which the single loop tunnel from T4 bifurcates into two tunnels serving the platforms in the central terminal area station.

The selected solution for the structure was a combination of a diaphragm wall structure to envelope the widest section of the turnouts needed, and sprayed-concrete-lined enlargements of short existing east and westbound lengths to accommodate the 'toe' of the turnouts. Sprayed-concrete-lined connections were also made to the bored extension tunnels

Fig. 14. A skid-steer mounted shotcrete robot was used to apply the Lasershell linings

Table 3. Shotcrete mix (water cement ratio 0.46)

Component	Source	Mass per m³: kg
Ordinary Portland cement	Castle Ketton	450
Micro silica (fluid)	Elchem	70
6 mm granite aggregate	Foster Yeoman	450
Dust (sand)	Foster Yeoman	1120
Water	-	170
Plasticiser (Glenium 51)	MBT	6
Stabiliser (Delvocrete)	MBT	1·1
Steel fibres	Dramix	30
Polypropylene fibres	Fibrin	1

Table 4. Key characteristics of pre-cast concrete segments

Characteristic	Project			
	Airside road tunnel	Storm water outfall tunnel	Piccadilly line extension	Heathrow Express extension
Segment type	Bolted	Expanded wedge block	Expanded wedge block	Expanded wedge block
Internal diameter: mm	8100	2910	4500	5675
Segment thickness: mm	350	220	150	220
No. of segments	8 (inclusive of 1 key)	7 (inclusive of 1 key)	8 (inclusive of 1 key)	10 (inclusive of 1 key)
Fibre reinforcement: kg/m³	None	30	30	30
Segment width: mm	1700	1000	1000	1000
Bar reinforcement: kg/m³	80	No	No	No
Polypropylene fibres	Yes	Yes	Yes	Yes
Concrete strength: N/mm²	60	60	60	60

from T5 (Fig. 15). The detailed construction aspects of the sprayed-concrete-lined tunnels are described by Stärk and Jäger.[13]

The structures were analysed for the temporary situation, including construction sequence loading – especially excavation and strut installation – as well as permanent load conditions.[14] Staadpro was used for the analysis of the box slabs and headwall structures in the temporary and permanent load conditions, while Flac3D was used for construction of the box structure and also to predict ground movements. Following construction, the site was to be returned for use as an aircraft stand so the roof slab was also designed to carry live loads from aircraft (typically Boeing 474-400s).

The tragic Nicoll Highway Collapse occurred in Singapore[15] around the time the junction box design work was completed. In line with its risk-management approach, the project team carried out a comprehensive review of the design and proposed construction methodology against the interim recommendations made by the Singapore Ministry of Manpower. The conclusion of this review was that the engineering aspects associated with the significant recommendations had already been considered and addressed.

Airside road tunnel portals

The airside road tunnel ports are formed by piled retaining walls.[3,4] The design was in accordance with the recommendations of CIRIA report 104.[16]

Implementation of the design was advanced by the use of the observational method to improve the execution of the design. This approach enables the team to gain time and cost advantages for the project.

Performance

TBM availability and programme durations

Detailed records were collected for each work shift, particularly of activity durations. The main purpose of this was to understand the distribution of time usage and analyse areas of downtime to allow continuous improvement and improved efficiency to be achieved. Table 5 summarises the key TBM availability and performance records.

The second drives for the Heathrow Express extension and airside road tunnels were more successful in terms of productivity. In contrast, the first Piccadilly line extension drive was the most successful. The Piccadilly TBM had the better overall reliability, with breakdowns related to the machine being about 5%. The breakdowns were limited to shove-ram hydraulic circuitry and chain conveyor drive motor failures.

Specific to the Heathrow Express extension TBM were mechanical issues related to the chain conveyor and to the 'flipper' mechanism on the cutting boom. On the airside road tunnel TBM, mechanical and electrical issues occurred over several component areas. The significant downtimes for the storm water outfall tunnel TBM consisted of head blockages in the initial section of the drive, and mechanical and electrical issues related to the segment erector vacuum pad.

Given that the Piccadilly line extension, Heathrow Express extension and airside road tunnels used the same TBM crew, it was reasonable to assume that productivity/performance variations were a direct result of TBM availability and performance (i.e. less downtime, increased productivity).

Ground movements

The tunnels crossed beneath various Heathrow airport infrastructure, including one of the runways, aircraft taxiways and stands, buildings, strategic fuel and fire mains and two existing underground railway tunnels. Despite having appropriate tunnelling methods, ground movement was inevitable and the methods used to determine the predicted ground movement is detailed in Pound *et al.*[17]

A holistic approach to the control of ground movement was developed covering the full spectrum, from selection of tunnel excavation method through to a robust geotechnical and TBM data-output monitoring-and-review cycles. This process did not prevent ground movement but provided the team with a robust framework

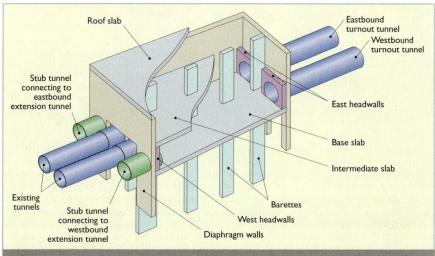

Fig. 15. Graphic of the junction box for the Piccadilly line extension tunnel[13]

Project	Tunnel drive	TBM activity breakdown									Duration and output			
		Production: %	Production (service extensions): %	Shift change/meal break: %	TBM delays: %	Grout delays: %	Survey delays: %	Conveyor delays: %	Other delays: %	TBM availability: %	Overall duration (days)	Average daily production: m	Best 24 h production: m	Best weekly production rate: m
Airside road tunnel	Eastbound (1st drive)	48	12	0	17	9	1	8		60	148	8.5	20.4	136
	Westbound (2nd drive)	53	23	0	6	4	0	6	10	76	102	12.6	20.4	134
Storm water outfall tunnel	N/A				8					79	117	38.2	72	459
Piccadilly line extension	Eastbound (1st drive)	54	21	12	4	N/A	4	2	3	87	53	26.9	50	258
	Westbound (2nd drive)	59	5	14	6	N/A	7	7	2	78	76	18.7	56	227
Heathrow Express extension	Downline (1st drive)	55	9	14	12	N/A	2	3	3	79	91	15.5	38	140
	Upline (2nd drive)	63	16	5	9	N/A	0	0	7	83	75	23.8	46	204

Table 5. TBM availability and performance

in which to manage the consequence with third parties. While the emphasis was on cause-control measures, detailed emergency and recovery plans were prepared in partnership with third parties to address the prevention of the occurrence of consequence escalation.

Table 6 summarises the typical and maximum settlement recorded for each project. The impact of all tunnels on the ground surface has been in the majority of cases within predictions. The key exceptions were the first 200 m of the Piccadilly line and Heathrow Express extension drives. The reason for this was that this area had been recently backfilled and the consolidation effects of the backfill were considerably greater than surface settlement caused by the tunnels. In the case of the Heathrow Express extension drives, the gathering of meaningful data for the initial stage was abandoned because of the ongoing earthworks activity above the tunnel alignments.

The surface-monitoring data for the airside road tunnel show that ground movements were kept within the maximum range and in most areas the actual ground movement was in close correlation to predictions.

In summary, the ground movement for over 36% of the drive length was between –5 mm and +5 mm and for 90% of the drive the movement was in the range of –10 mm and +10 mm. There was no significant movement recorded greater than 20 mm. In contrast to the other tunnels, heave was observed for the airside road tunnel – this occurred as a result of positive face pressure.

In addition to the control of global settlement, the significant challenge for the Piccadilly line and Heathrow Express tunnels was crossing beneath the existing Piccadilly line railway tunnel.

The movement recorded was about 20 mm for the Heathrow Express drives (5·75 m diameter passing 5 m beneath the existing and the tunnel) and 12 mm for the Piccadilly line drives (4·5 m diameter passing 4 m beneath). In both cases, the actual movement was within 5% of the predicted value and, in all cases, tunnel construction proceeded without affecting the safe normal service of the existing railway infrastructure and that of the airport.

Conclusion

A total of 14 km of incident-free tunnel infrastructure was delivered to support Terminal 5. The key factors in the success of this delivery were the implementation of the T5 agreement by a committed integrated team.

The implementation of the principles of the agreement by the team enabled the projects to be delivered within cost and time targets, and to develop innovative methods to enhance safety.

Acknowledgements

The author thanks the many individuals and companies who were fully committed to the principles of the T5 agreement – trust, commitment and teamwork – without it none of the successes would have been achieved.

Notwithstanding the commitment of individuals, the author acknowledges the courage of BAA to implement the T5 agreement and head of rail and tunnels Ian Fugeman's relentless commitment and support, which allowed professionals the opportunity to achieve exceptional performance.

Table 6. Summary of ground movements for each tunnel

Tunnel	Excavated diameter: m	Typical settlement: mm	Maximum settlement: mm	Volume loss: %	Notes
Airside road tunnel	9·16	10	20	0·35	Heave also observed
Storm water outfall tunnel	3·35	6-8	–	1	
Piccadilly line extension	4·81	10	14	1 to 1·5	
Heathrow Express extension	6·125	20	33	1·5 to 2	Maximum settlement coincided with blocky ground conditions

References

1. EDDIE C. and NEUMANN C. Lasershell leads the way for SCL tunnels. *Tunnels and Tunnelling International*, June 2003.
2. HILAR M. and THOMAS A. Tunnels construction under the Heathrow Airport. *Tunnel*, 2005, **14**, 17–23.
3. HITCHCOCK A. Elimination of temporary propping using the observational method on the Heathrow Airside Road Tunnel (ART) project. *Ground Engineering*, 2003, **36**, No. 5, 30–33.
4. POWDERHAM A. J. From major to minor – innovation and managing risk on small projects. *13th Danube–European Conference, Active Geotechnical Design in Infrastructure Development, ISSMGE*, 2006, Ljubljana.
5. WILLIAMS I. and AUDUREAU J. L. Large diameter TBM for tunnelling shallow tunnels beneath Heathrow Airport, UK. *54th Geomechanics Colloquy*, Austria, 2005.
6. LISTER B. Heathrow Terminal 5: enhancing environmental sustainability. *Proceedings of the Institution of Civil Engineers, Civil Engineering*, 2008, **161**, Special issue—Heathrow Airport Terminal 5, May, 21–24.
7. LANCE G. and ANDERSON J. *The Risk to Third Parties from Bored Tunnelling in Soft Ground*. HMSO, London, 2006, Health & Safety Executive Research Report 453.
8. HEALTH & SAFETY EXECUTIVE. *The Collapse of NATM tunnels at Heathrow Airport. A Report on the Investigation by the Health and Safety Executive into the Collapse of New Austrian Tunnelling Method (NATM) Tunnels at the Central Terminal area of Heathrow Airport on 20/21 October 1994*. HSE Books, London, 2000.
9. THE ASSOCIATION OF BRITISH INSURERS and THE BRITISH TUNNELLING SOCIETY. *Joint Code of Practice for the Procurement, Design and Construction of Tunnels and Associated Underground Structures in the UK*. Thomas Telford, London, 2003.
10. LONDON UNDERGROUND LTD. *Engineering Standard 1008: Assurance*. LUL, London.
11. EVANS M. Heathrow Terminal 5: health and safety leadership. *Proceedings of the Institution of Civil Engineers, Civil Engineering*, 2008, **161**, Special issue—Heathrow Airport Terminal 5, May, 16–20.
12. WILLIAMS I., NUEMANN C., JAGER J. and FALKNER L. Innovative shotcrete tunnelling for London Heathrow's new Terminal 5. *53rd Geomechanics Colloquy and Austrian Tunnelling Day*, Salzburg, Austria, 2004.
13. STARK A. and JAGER J. London Heathrow Terminal 5 – construction and monitoring of Piccadilly extension junction. *Underground Space – 4th Dimension of Metropolises*, Prague, 2007, pp. 1341-1347
14. JAGER J. and STARK A. Deformation prediction for tunnels at Piccadilly line extension junction in London Heathrow – an engineering approach. *Underground Space – 4th Dimension of Metropolises*, Prague, 2007, pp. 463–469
15. MAGNUS R., CEE ING T. and LAU J. M. *Report in the Incident at the MRT Circle Line Worksite that led to the Collapse of the Nicoll Highway*. Ministry of Manpower, Singapore, 2005.
16. GABA A. R., SIMPSON B., POWRIE W. and BEADMAN D. R. *Embedded Retaining Walls – Guidance for Economic Design*. CIRIA, London, 2003, report 104.
17. POUND C., HSU Y. S. and WALKER G. R. Use of three-dimensional numerical methods to predict ground movements around tunnel boring machines. *Underground Construction*, 2003, 549–562.

What do you think?

If you would like to comment on this paper, please email up to 200 words to the editor at journals@ice.org.uk.

If you would like to write a paper of 2000 to 3500 words about your own experience in this or any related area of civil engineering, the editor will be happy to provide any help or advice you need.

Proceedings of ICE

Civil Engineering 161 May 2008
Pages 38–44 Paper 700045

doi: 10.1680/cien.2007.161.5.38

Keywords
airports; piles & piling; subsidence

Tim Dawson
BSc, CEng, MICE

is a project director at Mott
MacDonald

Kathiresapillai Lingham
BSc

is senior engineer at Mott
MacDonald

Roger Yenn
MSc, BSc, CEng, MICE

is a director at TPS Consult

Jim Beveridge
BSc, CEng, MICE, MIMMM

is foundation and ground
engineering director at Mott
MacDonald

Richard Moore
BEng, CEng, MICE

is a technical associate director at
TPS Consult

Matthew Prentice
BEng, CEng, MICE

is a project manager at Laing
O'Rourke

Heathrow Terminal 5: building substructures and pavements

This paper describes the design and construction of the vast piled basement structures for the three terminal buildings at London Heathrow's £4·3 billion Terminal 5 project, together with 1 million m² of associated aircraft pavements. The basements are up to 20 m deep and involved the excavation and reuse of 6·5 million m³ of gravel and clay. The aircraft pavement involved a number of innovations including development of a new high-strength concrete, which delivered a thinner construction and resulted in programme and environmental benefits.

London Heathrow airport's £4·3 billion Terminal 5 (T5) project involved extensive foundation and paving works.

The basement structures for the three terminal buildings—the core T5A terminal (Fig. 1) and satellites T5B and T5C—are up to 20 m deep. Their design mitigates the effects of buoyancy from the groundwater and heave of the underlying London Clay, and they are supported by large-diameter piles. Single piles are provided at column positions and the design, based on extensive testing and three-dimensional geotechnical modelling, accommodates strict limits on differential settlement between adjacent columns.

The associated earthworks comprised the excavation and reuse of 6·5 million m³ of superficial deposits, terrace gravel and London Clay. Detailed investigations, testing and analysis were required to ensure the stability of the deep excavations. The heave potential of the London Clay fill was mitigated through site conditioning developed during full-scale field trials.

The construction of the adjacent aircraft pavements also included a number of innovations, including the development of a new

high-strength concrete that delivered a thinner pavement and programme and environmental benefits. Unsealed joints were also introduced for the first time on BAA aircraft pavements.

The principles of standardisation, prefabrication and modularisation were embraced at every stage of the project. All bulk materials were brought in by rail to the nearby Colnbrook logistics centre, where prefabrication of reinforcement cages and rollmats was carried out.

Construction efficiency and quality were significantly improved by the integrated design and construction team. The organisation of the T5 substructures and civil works team is shown in Fig. 2.

Earthworks

The construction of T5 involved the excavation of approximately 6·5 million m³ of material and the compaction of some 7·2 million m³ within a contaminated, waterlogged site of only 1 km² in area. The site, formerly Thames Water's Perry Oaks sludge disposal works, was bisected by two man-made rivers that had to be diverted around the western perimeter of the airfield.[1]

Fig. 1. The basement of the core T5A terminal building is up to 20 m deep

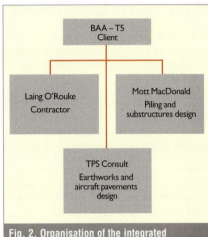

Fig. 2. Organisation of the integrated substructures and civil works project team

A series of site investigations was carried out prior to the commencement of works and in the early stages of the project. The typical ground profile is outlined in Table 1. The groundwater level was typically 3 m below existing ground level. In some locations the terrace gravel had been completely removed and either replaced with made ground or the top of the London Clay was left exposed.

The main source of soil contamination on the site was from sewage sludge found beneath the floor of some of the lagoons and within the bunds. Approximately 700 000 m³ of bund material contained a proportion of sludge. The older sewage sludge contained high levels of heavy metal contamination, reflecting more liberal discharge consents in the past. However, the organic content of the sludge tended to bind these metals in the soil and there was little evidence of leaching into the groundwater.

Dewatering

The terrace gravel required dewatering to enable construction to proceed. Over 3·3 km of cement–bentonite cut-off walls were installed to permit zonal dewatering of the site linked to the proposed construction sequence. Deep sumps were constructed around the site to reduce the initial groundwater level within the terrace gravel and to control groundwater from precipitation infiltration. Groundwater within the terrace gravel, containing elevated levels of ammonia from previous use of the site, was discharged to Mogden treatment works in Isleworth via the Bath Road sewer culvert at the north of the site. Two separate temporary lagoons were constructed to balance the flows from the dewatering system before discharge to the culvert.

Open-cut excavations

The project involved the construction of substantial underground structures for the three terminal buildings—T5A, T5B and T5C—requiring excavations of up to 20 m depth. The resulting excavation into London Clay was deeper than anything previously attempted. One of the early decisions was the choice of excavation strategy: retained vertical excavations or open-cut. Open-cut excavations increased the volume of earthworks and aggravated an already constrained site, but offered significant cost and programme savings and were the preferred solution.

Few published data were available on the short-term (up to 18 months) stability of deep excavations in London Clay. After extensive modelling and analysis, the profile illustrated in Figs 3 and 4 was adopted for the deep excavations. It was important that groundwater seepage and run-off was controlled to prevent softening of the clay, particularly at the gravel/clay interface and at the top of the slope. This was achieved through a system of engineered drainage.

An additional concern was the possibility of failure along sub-horizontal shear zones attributed to ancient tectonic movements, as had occurred at the nearby Prospect Park.[2] Although evidence gathered from the investigation phase suggested that similar tectonic shear zones could be encountered at T5, no attributable failures were observed.

Construction of the substructure for T5C at the eastern end of the site presented an unusual

Table 1. Ground conditions	
Stratum	**Elevation: m OD**
Terrace gravels	22·4 to 18·5
London Clay	18·5 to −32·0
Lambeth Group	Below −32·0

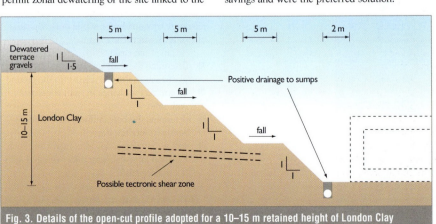

Fig. 3. Details of the open-cut profile adopted for a 10–15 m retained height of London Clay

Fig. 4. Piccadilly line extension launch chamber excavation showing benched profile

challenge as it involved significant excavation and stockpiling over the recently completed Piccadilly line and Heathrow Express extension tunnels. This demanded careful planning and control to avoid causing unacceptable distortions in the tunnels.

Slope monitoring was carried out throughout all works. This comprised inclinometer monitoring and in-situ suction measurements on some of the most critical slopes, combined with a rigorous regime of visual inspections by qualified personnel. A contingency plan was in place if monitoring showed that a failure surface was developing, in which case this surface would be interrupted with sheet piles. This eventuality never arose.

Table 2. Excavation volumes	
Material	**Approximate volume: million m³**
Superficial material	2·5
Terrace gravels	1·3
London Clay	2·7
Total	**6·5**

Fig. 5. Major sheet-piled walls up to 18 m retained height were needed next to the twin rivers so that excavation and backfilling could take place prior to their diversion

Fig. 6. Around 4600 m² of temporary reinforced soil walls up to 18 m high were also used where space restrictions and logistical and programme constraints dictated

Retained excavations

It was not possible to rely solely on open-excavation and some vertical support was necessary where space restrictions or logistical and programme constraints dictated. In particular, the need to complete excavations and backfilling prior to diversion of the twin rivers required the use of sheet-pile retaining walls and reinforced soil walls (see Figs 5 and 6). Approximately 9000 m² of temporary sheet-pile wall was installed (cantilever and anchored) over the course of the project, with a maximum retained height of 18 m. About 4600 m² of temporary reinforced soil walls were also constructed up to a maximum height of 18 m.

Reusing spoil and fill

A primary environmental target for T5 was to maximise the reuse of material generated within the site. The approximate volumes of materials excavated are shown in Table 2. A substantial volume of London Clay was generated from the deep excavations and the tunnel bores. However, such material placed as fill at shallow depth has the potential for significant swelling, which could be detrimental to substructures or overlying pavements. Testing showed that the potential for swelling increased noticeably at effective stresses below 50–100 kPa,[3] the range of stresses applicable to fill on T5. The following solutions were adopted.

- Site-won gravels were used as backfill to substructures to minimise lateral swelling pressures and reduce lateral earth pressures on the walls.
- Where London Clay was used as a general fill under aircraft pavements, it was placed in a controlled manner. This involved leaving a period of time between placement of the material and construction of the pavement where possible, and conditioning the material by the addition of water to reduce its heave potential.
- A full-scale earthworks field trial was carried out at an early stage in the project. This yielded a greater understanding of the behaviour of compacted London Clay and allowed development of a cohesive superficial material and verification of the levels of compaction achievable using a compaction method specification. The trial also investigated the use of lime and cement to stabilise and modify the earthworks materials.

Piling

The terminal building structures are supported by bored piles up to 2·1 m maximum diameter founded in London Clay and extending to a maximum depth of 35 m below the lowest basement level. Drained under-floor cavities were used to reduce ground loading on the base slab by allowing unrestrained heave of the underlying clay. Heave was predicted to exceed 100 mm to a depth of at least 18 m below excavation level and there was also concern that long-term swelling could lead to softening of the clay, thus reducing the long-term shaft load capacity of the piles.

The design of the piles was dictated by the need to prevent significant differential movement between adjacent piles and the desire for economy in overall pile design. Also, to speed up construction, the design was developed using only one pile per column, negating the need for pile caps. The maximum pile load is 3500 t.

The initial pile design parameters suggested that a significant number of under-reamed piles with under-ream diameters of up to 6·2 m would be required. This was undesirable from a construction and programme perspective, and one objective was therefore to reduce these to a minimum. Base stiffness and shaft resistance were thus important considerations.

Pile load tests

The groundwater table is located within the terrace gravels at an elevation of 21·8 m. To maximise economies in the pile design, a suite of pile tests was carried out to

- determine the shaft resistance at large displacement of the pile relative to the ground
- determine the change in shaft resistance with time
- determine the base stiffness and ultimate resistance of the piles
- ensure the practicality of constructing such long and large-diameter piles with and without under-reams.

To assess the shaft resistance, four 1·05 m diameter, straight-shafted in-situ concrete piles with soft toes were installed to a depth of

Proceedings of ICE

Civil Engineering 161 May 2008
Pages 45–53 Paper 700051

doi: 10.1680/cien.2007.161.5.45

Keywords
airports; buildings, structures & design

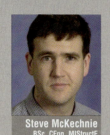

Steve McKechnie
BSc, CEng, MIStructE

is associate director at Arup

Dervilla Mitchell
BE, CEng, MIEI, MICE, FREng

is a director at Arup

William Frankland
CEng, FICE

is engineering manager at
Laing O'Rourke/BAA

Maurice Drake
MA, CEng, MICE, MIStructE

is an associate director at Arup

Institution of Civil Engineers

CIVIL ENGINEERING

Heathrow Terminal 5: terminals T5A and T5B

This paper describes the design and construction of the two terminal buildings in phase 1 of the £4·3 billion Terminal 5 development at London's Heathrow airport. A total of 40 000 t of steel was used to create 280 000 m² of space in the dramatic main terminal building, T5A, including forming a clear span roof of 156 m by 396 m to enclose it. In addition to high visibility for passengers, the design provides maximum flexibility for future modifications. In the first of two satellite buildings, T5B, 600 000 m² of post-tensioned flat slabs were cast using minimal amounts of formwork and site labour. Building these vast structures at the world's busiest international airport also meant all construction operations had to be undertaken within a highly restricted space and with no cranes allowed above roof level.

Terminal 5 (T5) at London's Heathrow airport opened to passengers on 27 March 2008. The two terminals that make up phase 1 of the £4·3 billion development—T5A and T5B—were designed to be highly functional airport buildings but with a special ingredient that would enable owner BAA to realise its vision of T5 as 'the world's most successful airport development.'

A total of 40 000 t of steel was used to create 280 000 m² of space in the main terminal building, T5A, and form a clear span roof 156 m × 396 m to enclose it (Fig. 1). This dramatic structure is the focal point of T5 and was required to provide a welcoming, friendly and efficient interchange for departing, arriving and transferring passengers.

The architecture has achieved a transparent quality, with views out to the airfield, which assists passengers with intuitive way-finding to and from their aircraft. It also provides the operator with maximum flexibility to accommodate future modifications to the internal building arrangements.

In the first of two satellite buildings, T5B, 600 000 m² of post-tensioned flat slabs were cast, using minimal amounts of formwork and

site labour. Both building designs were also informed by the constraints of building on the world's busiest international airport, where site space for construction operations and storage is highly restricted and the low radar ceiling limits any use of cranes above roof level.

Figure 2 is a simplified cross-section showing how T5 operates. Departing passengers arrive in T5A on bridges from the rail interchange, forecourt and car park, and then check in and clear security on the upper level. From there they either proceed to the middle level of T5A, or via a tracked transit system to T5B, for gate seating and retail space. Arriving passengers use the lower level of T5A to clear immigration, collect their bags and move out towards surface transport for their onward journey.

T5A

T5 was built on one of the world's busiest international airports, adjacent to one of Europe's busiest motorway networks. Site space was at a premium: the T5A site was surrounded by a 4000 space car park, the original line of the twin rivers and various substructures. The exter-

Fig. 1. Main terminal building T5A has a transparent quality, with views out to the airfield which assists passengers with intuitive way-finding to and from their aircraft

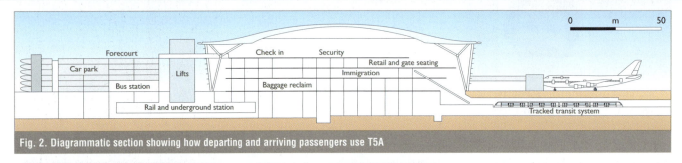

Fig. 2. Diagrammatic section showing how departing and arriving passengers use T5A

Fig. 3. Open-plan check-in and security areas are directly under T5A's 156 m span, 390 m roof enclosure

Table 1. Key members of the terminal buildings project team

T5 role	Organisation
Client	BAA
Architect	Richard Rogers Partnership
Architectural services and design management at scheme stage	HOK
Architectural services on T5B	Pascall and Watson
Structural engineer for above-ground structures	Arup
Structural engineer for substructure	Mott MacDonald
Structural steelwork for roofs and facades	Watson Steel Structures Ltd
Structural steelwork for superstructure	Severfield Rowen plc
Concrete works and team management for T5A roof	Laing O'Rourke plc
Roof cladding	Hathaway Roofing Ltd
Façades	Lindner Schmidlin Facades (previously Schmidlin)
Temporary works for T5A strand-jacking and abutment assembly	Rolton Group
Rafter post-tensioning	Bridon
Strand-jacking	PSC Fagiloi
Mechanical and electrical services	Amec

nal envelope build was on the critical path for the opening of the development and the 'radar ceiling' for the airport meant that cranes were not allowed higher than 2 m above the highest point of the building without special permission.

BAA brought together a world class team of construction professionals, working under a partnering contract (Table 1). They developed construction solutions, some of which were quite specialised, to minimise the impact of the site and time constraints and to ensure that the safety record for the project was significantly better than industry practice.

The structural concept for T5A is essentially simple: a three-storey steel-frame supports 280 000 m² of composite floor slabs above a basement that is up to five levels deep. All this is enclosed by a clear-spanning roof and façade structure.

Single-span roof

The roof has a span of 156 m and is 396 m long. It is supported by 22 pairs of 914 mm diameter steel legs that reach down to apron level, leaving dramatic full-height spaces just inside the façades. These spaces form the main routes for passenger vertical circulation to and from the gates.

The span is formed from steel box girders at 18 m centres, 800 mm wide and up to 3·8 m deep. These are tied at high level by pairs of 115 mm diameter pre-stressed locked-coil strand cables. Steel arms 914 mm diameter

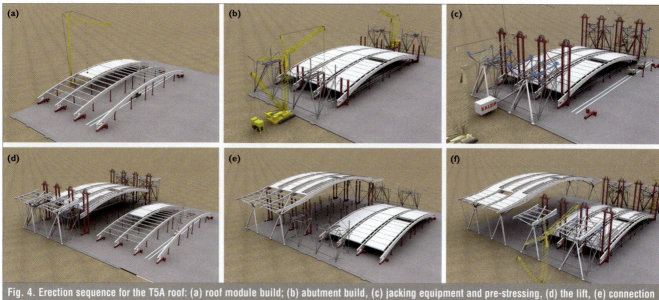

Fig. 4. Erection sequence for the T5A roof: (a) roof module build; (b) abutment build, (c) jacking equipment and pre-stressing, (d) the lift, (e) connection and load transfer, (f) cladding and infill

reach up from the tops of the legs to support the rafters, and solid steel tie-down straps from the rafter ends complete the thee-dimensional hybrid portal frame structure.

The team examined a number of different structural arrangements for the roof during design development. Shorter span options used less steel, but the single clear span offered advantages that outweighed the attractions of a mere saving in tonnage.

■ The clear span structure allowed the roof and façades of the building to be erected and clad very quickly, which led to an early watertight date and an early start to the mechanical and electrical packages and fit-out. Moreover, the internal structure was constructed in a semi-indoors environment, improving build quality and reducing programme delay from bad weather.

■ The clear span roof could be assembled close to ground level and lifted into position with strand-jacks. This not only reduced the risks of working at height for both the steel erectors and the roofing installers but it also allowed virtually the whole operation to be carried out by cranes that were below the radar ceiling. If the roof had been built on top of the superstructure, then every lift of steel and cladding materials would have needed special permission from the control tower.

■ In the multi-span schemes, BAA's space planners found that the presence of roof supports placed a constraint on the layout of check-in, security and retail areas. By contrast they valued the truly open-plan space that the clear span gave them (Fig. 3). Furthermore, future layout changes in these

areas will be simple because they will not involve any structural works. Disruption to passengers, and thus BAA's business, will be kept to a minimum.

■ The roof acts as a visually unifying element for the terminal: wherever users are in the terminal they can look upwards and immediately get a feeling for where they are, both in the building, and in the world. The lines of the structure and roofing are deliberately simple and clean to impart a feeling of calm and purpose to the space. This has a direct effect on the efficient running of the terminal and its attractiveness to passengers.

Programme and build sequence

Work on the basements started towards the latter part of 2002 involving bulk excavation, piling, reinforced concrete (RC) walls and slabs and heavy structural steelwork with composite RC decks.[1]

The roof structure started in October 2003 and was constructed in six phases: five of 54 m length and one of 18 m. The central arched section of each phase was assembled, partially clad and pre-stressed at ground level. Temporary

works frames were used to position the abutment steel accurately. The centre section was then jacked 30 m vertically into position and bolted to the abutment steel, after which the final roof skin was installed. Once each phase of roof was complete, the temporary works frames were rolled north by 54 m ready for the next phase. After the next phase was up, purlins were inserted into the infill and the roof cladding was installed; all working from ground level using mobile cranes. Fig. 4 shows the construction sequence.

Façade steel was then erected and the super-structure frame was built under the completed roof. Work continued on completion of roof details, in particular the insertion of the services chimneys and downspouts, until the latter part of 2006.

Figure 5 shows the programme for the construction of the main components of T5A described in the current paper. An eight-week cycle was achieved for the erection of each roof module, giving an overall installation period of just 15 months to erect some 18 000 t of roof steelwork.

The following paragraphs describe the roof

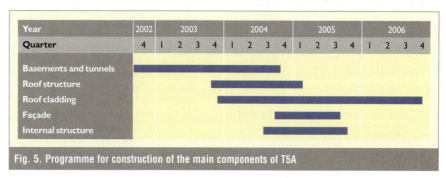

Year	2002	2003				2004				2005				2006			
Quarter	4	1	2	3	4	1	2	3	4	1	2	3	4	1	2	3	4
Basements and tunnels																	
Roof structure																	
Roof cladding																	
Façade																	
Internal structure																	

Fig. 5. Programme for construction of the main components of T5A

Fig. 6. Abutment temporary work support frameworks positioned next to the first roof module; they were moved horizontally along the apron slab as the roof erection progressed

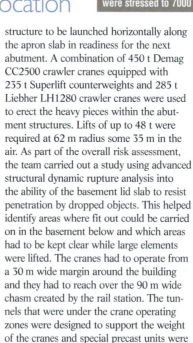

Fig. 7. First 2000 t roof module being lifted 30 m into position by four sets of tower-mounted, computer-controlled strand-jacks. The horizontal cable ties were stressed to 7000 kN once the module ends had cleared the ground

the level of the roof apex was allowed to vary by up to 150 mm since there are no critical interfaces in this location

and superstructure build in more detail.

- *Roof module build* (Fig. 4(a)). Twin 70 t Kobelco CKE 700 crawler cranes and an 80 t Grove GMK 4080 mobile crane were used to erect the central rafter sections, bracing and purlins. The temporary props and strand-jack towers were connected directly to the heads of columns in the basement below and temporary spreader beams transferred loads from the craneage onto substructure beam lines. Rafters were supplied in 9 m and 18 m sections with simple direct-bearing splice connections, so very little site labour was required. A long-radius Spierings SK 598-AT5 hydraulic folding crane was used for erection of the roofing cassettes and other roofing material. Around 80% of the roofing materials were supplied to the site in the prefabricated cassettes: 3 m wide and spanning 6·25 m between purlins. Kalzip panels were rolled on site and placed on the roof modules ready for installation after the lift.

- *Abutment build* (Fig. 4(b)). The abutment temporary support framework had to support the abutment permanent works and provide a means of adjustment to achieve the correct geometry prior to connection to the roof module. This was done through six vertical and horizontal jacking points. Twin longitudinal vertical trusses were also incorporated into the abutment temporary works framework (Fig. 6) to enable the

structure to be launched horizontally along the apron slab in readiness for the next abutment. A combination of 450 t Demag CC2500 crawler cranes equipped with 235 t Superlift counterweights and 285 t Liebher LH1280 crawler cranes were used to erect the heavy pieces within the abutment structures. Lifts of up to 48 t were required at 62 m radius some 35 m in the air. As part of the overall risk assessment, the team carried out a study using advanced structural dynamic rupture analysis into the ability of the basement lid slab to resist penetration by dropped objects. This helped identify areas where fit out could be carried on in the basement below and which areas had to be kept clear while large elements were lifted. The cranes had to operate from a 30 m wide margin around the building and they had to reach over the 90 m wide chasm created by the rail station. The tunnels that were under the crane operating zones were designed to support the weight of the cranes and special precast units were used to span over the air-intake shafts, which occur at numerous intervals.

- *Jacking equipment and pre-stressing* (Fig. 4(c)). The central section was raised using four sets of strand-jacks mounted on jacking towers at either end of the central section (Fig. 7). Strand-jacking equipment consisted of eight pairs of 180 t strand-jack rams equipped with strand re-coilers. Each lifting cable was composed of twelve

18 mm Dyform strands which ran from the jacks to dead-end anchorages either side of a lifting beam cradling the ends of each rafter. In order for the central arch section modules to span the 117 m between jacking towers, the 115 mm high-tie cables had to be pre-stressed to 7000 kN per pair. The first stage was to lift the ends of the arches 200 mm clear of the props. At this stage the centre of the arches did not move. The high-ties were then stressed using the cable-tensioning jacks that were connected between the tie connection point and the cable termination. The pre-stress in the cables caused an upward bending moment in the rafters, which deflected upwards, lifting themselves off the central supports and transferring their full 2000 t weight to the strand-jack system.

- *The lift* (Fig. 4(d)). The strand-jacks were then used to raise the central section into place 30 m above the ground floor slab. Strand-jacking was coordinated via a computer-controlled system, which enabled automatic jack stroking to synchronise ram movement within prescribed lifting control criteria, supplemented by an external survey-based system.

- *Connection and load* (Fig. 4(e)). The abutment structures were adjusted to the required geometry and then connected to the central section with proprietary high-strength friction-grip bolts in the main rafter splice (Fig. 8). Removal of the tem-

Fig. 8. The abutment structures were spliced to the roof module rafters at each end with high strength friction-grip bolts before strand-jacks were released

porary works was carefully choreographed to ensure load was transferred to the permanent works without overloading any components.

- *Cladding and infill* (Fig. 4(f)). Much of the installation of roof cladding, glazing, and services was carried out manually using materials that had been lifted with the roof module. The abutment temporary works were lowered and moved north using horizontal jacks and Hillman rollers ready for the next abutment build. Infill sections between the modules were then installed.
- *Façade and superstructure.* Façade steel was erected using cranes working from the 30 m margin. The internal superstructure steel was erected generally traditionally, except that cranes were operating under a roof that had already been erected. The fabrication and delivery of batches of steel was programmed to ensure that superstructure erection zones coordinated with areas of roof that had been completed.

Managing positional accuracy

It is inevitable that the as-built position of any civil engineering works will deviate to some extent from the position shown on the design drawings. Under a traditional contract, this often leads to recriminations and remedial works and can delay the rest of the project. The T5A team decided to do all it could to eliminate such potential problems by designing adjustment into the structural frame of the roof.

The first step was to identify the locations on the frame where positional accuracy was important for interfacing packages, and to quantify a desirable level of accuracy for those points. Arup carried out an analysis of the effect that individual dimensional deviations of the steel components would have on the geometry of the overall frame, and used a statistical analysis to calculate the probable combined effect of all those deviations. It was then possible to quantify the adjustment that would be required to allow the frame to be built with an acceptable geometry, and the connections were designed accordingly.

Locations on the frame where accuracy was not critical for fit-up or following trades were allowed a greater permissible deviation. For instance, the level of the eaves was controlled to within 15 mm because this was critical for the fit of the façade glazing, but the level of the roof apex was allowed to vary by up to 150 mm since there are no critical interfaces in this location.

Abutment first-run study

BAA was aware of the risk to the whole programme if the roof construction became delayed. It therefore accepted the project team's proposal to assemble one complete abutment structure at Dalton in North Yorkshire prior to its erection on site.

The so-called 'abutment first-run study' involved the actual components and methods which would be used on site, and was carried out by the same team of steel erectors. Its pur-

pose was to 'practice, prove and streamline' the assembly method. The study took place some eight months before actual start on site. The crew became familiar with the intricacies of the installation methodology and the use of hydraulic jacks provided for positional adjustment of the abutment steelwork.

Over 100 improvements to erection methodology and safety were identified and incorporated. In so doing, the exercise potentially saved in excess of 100 site days—which could have cost the client approximately £4 million. The study also included trialling of the roof cladding, the façade structure, glazing installation method and elements of the feature lighting.

Structural behaviour during load transfer

Analysis suggested that the steelwork would drop by 87 mm when the temporary works were removed; that the legs would shorten by 3 mm; and that the horizontal ties that resolve the forces at the nodes where the abutment steel meets the rafters would extend by 6 mm. These dimensional changes are simply the consequence of stress in the steelwork from its self-weight and the weight of roofing materials. Their overall effect could, however, have had a significant impact on the final shape of the roof, so the team decided to pre-set the steelwork dimensions to counteract the majority of these dead-load deflections.

The first roof module was thus lifted to 100 mm above the position shown on the drawings so that dead-load deflection on de-propping would carry it into the 'correct' position. The site team took regular measurements of the steelwork during erection as a risk control measure and to ensure that the permissible deviations were being achieved. These measurements generally showed that the steel was behaving as expected but, when it came to the final stage of the load transfer in the first lift, they showed that the central module had moved downwards by 182 mm rather than the 87 mm that had been anticipated. This had not been predicted by the analysis.

The analysis had, of course, only taken account of the deflections owing to elastic behaviour of the steel. The connections in the roof frame either rely on direct bearing of machined steel faces or on the bearing of bolts in clearance holes. What the analysis had not taken account of was the movement that occurred while all these faces moved into bearing.

The evidence of the first load transfer showed that the roof moved by 95 mm as the connections came into bearing. Subsequent lifts were therefore pre-set by lifting them an extra 90 mm above the position shown on the drawings. The main rafter splices were partially connected and the roof was lowered until the gaps at the bearing faces had started to close. It was possible to tell when the gaps had started to close

Fig. 9. On the western and eastern elevations of T5A, the façade is supported by 400 mm by 200 mm elliptical hollow sections that span 18 m horizontally between the roof tie-down straps

Fig. 10. Pin-ended struts at the top of the T5A gable façades allow 150 mm vertical movement and in-plane horizontal movement between the gable and roof

Fig. 11. Glazing on the T5A western elevation opposite the multi-storey car park is designed to resist a vehicle bomb blast

because the hydraulic pressure readings in the strand-jacks started to drop, indicating that the permanent structure was taking a significant proportion of the load. This turned out to be quite a gradual process, but the downward deflection of the roof was close to 95 mm, which was consistent with the team's hypothesis about closing the gaps at the connections.

The splices were connected at bottom flange level only at this stage so they could carry the axial force needed to close the connection gaps, but they would not allow secondary bending moments to build up in the rafters. For the final stage, the rafter splices were fully connected and the propping was removed completely. The roof moved downwards by a further 87 mm or so.

Façade

A key part of the passenger experience in T5A is the ability to look out at the airfield and aircraft and get a taste of the excitement of air travel. The façades are therefore fully glazed, and the design team strove to minimise the intrusion of vertical structural elements into oblique views through them.

On the western and eastern elevations, the façade is supported by 400 mm by 200 mm elliptical hollow sections that span 18 m horizontally between the roof tie-down straps (Fig. 9). The weight of glass and steel is carried to apron level by a series of 139 mm diameter steel props. The roof tie-down straps are a part of the roof structure: they run vertically, carrying tensions of up to 9000 kN. They take advantage of the stiffening effect of this tension to reduce the bending moments and deflections from façade wind loads, enabling them to be slimmer and less obtrusive than a conventional façade support.

The façades on the gable ends of the building each consist of a simple grid of steel that carries gravity loads down to apron level and resists wind loads by spanning vertically up to the underside of the roof. There is a joint at the head of the gable façade to allow a vertical movement of 150 mm and an in-plane horizontal movement of 100 mm between the façade and the roof while still carrying wind load in the out-of-plane direction (Fig. 10).

On the western elevation, where the terminal building faces the public car park, the glazing system is designed to be resistant to blast (Fig. 11). A vehicle bomb parked in the car park could break the glass but the interlayer between the glass laminations would prevent the broken glass from flying into the building and causing injury.

The pressure wave from a bomb blast would be very intense when it hit the glass, in excess of 40 kN/m², but would have a very short duration. Within a few milliseconds it would have been followed by a negative pressure wave. The load 'felt' by the supporting structure varies with the inertia of that structure. The slower the structure is to react, the less force it feels. The overall roof structure would be unaffected by the blast, but the façade support steel has been designed with enhanced connection capacities to ensure that the full plastic capacity of the steels can be mobilised.

The façade itself is a unitised system, and is fixed to the frame by brackets that allow up to 25 mm adjustment in any direction. The façade supplier used a purpose-built mechanised handler on a rail-mounted A-frame to install the glazing.

Internal structure

The steel/composite internal structure has a grid of up to 18 m by 18 m to allow flexibility in the planning of the baggage systems and retail areas. A grid of 18 m by 9 m is used for economy in less critical areas (Fig. 12). The storey height is typically 6 m, which allows structural and services zones to be completely separate to simplify services installation and reduce the risk of clashes. The 150 mm thick normal-weight concrete floor slabs are cast on, and act compositely with, 'Holorib' re-entrant metal decking.

The internal structure has two full movement joints that divide it into three structurally separate zones, each one measuring around 150 m by 150 m. It is stabilised by bracing in stair and lift core walls. The building contains significant areas of retail and so the floor beams and slabs are designed to allow some flexibility for areas of slab to be removed.

The basement suspended slabs are framed in steelwork, with the beams at ground floor level sized to provide a tie across the width of the building to restrain the feet of the roof columns.[1]

T5B

T5B is approximately 450 m long and 51 m wide (Fig. 13). It comprises a clear-span steel roof on two floors of post-tensioned flat-slab superstructure on a conventionally RC substructure. Its proximity to the much wider and taller T5A means that it is easy to forget how large T5B is: it is larger in plan area, for example, than Terminal 4. If it had been built on its own without the rest of T5, it would have been BAA's largest project for the last ten years.

T5B's design above ground level was influenced very heavily by the need to keep within the height limitations imposed by the planning inquiry. A further requirement was the need for flexibility during construction and subsequent use of the building. In 2000, during the early stages of design, the retail architects stated that the average lifetime of a retail project from concept through construction and use and on to demolition or complete refurbishment was seven years, so any scheme agreed then would be completely out of date by the time the build-

ing opened in 2008. This discouraged the tight integration of structural and services zones, and meant that the structural system chosen had to be capable of dealing with significant modifications in the future. This, together with the space requirements for services and the brief to include a mezzanine floor for commercially important passenger lounges, led to a very tight zone available for the structure.

The third major influence on the form of the structure was the architectural desire to cantilever the second floor over 3 m beyond the perimeter of the first floor.

The constraints led to the final superstructure solution of post-tensioned flat slab. This is a well-known option for offices but is unusual for a terminal. The system scores highly for overall depth. It also scores highly for its capability to

Fig. 12. Columns in the T5A baggage reclaim area are on the standard 9 m by 18 m grid: storey height is 6 m to allow maximum flexibility of services in the ceiling areas

Fig. 13. The 450 m long, 51 m wide T5B is bigger in plan area than Terminal 4

Fig. 14. The two above-ground floors of T5B are made from pre-stressed flat concrete slabs with grouted tendons at 1 m centres; this allows more flexibility for new service holes

Fig. 15. T5B steel roof trees were welded in jigs on site and then delivered to the building and craned into position

cantilever out beyond the line of external columns; indeed, it is a positive advantage with a post-tensioned flat slab to have a small cantilever, since ideally the tendons will be near the top of the slab on the support line but near the centre of the slab at the end anchorage. Having a cantilever enables this to happen. Post-tensioned flat slabs are not well known for their ability to be modified after construction, and this was the subject of further investigation.

Design for flexibility

Future modifications can be divided into two groups: those that do not cut tendons and those that do. Tendon ducts are generally spaced at much wider centres than conventional reinforcement, so it is easier to find suitable locations for small hole than it is for a conventional RC slab. Typically every 9 m bay had nine ducts, giving an average spacing of about 1 m (Fig. 14). Around column heads, shear reinforcement was also provided; this was, however, standardised for ease of construction, meaning most locations

were over-provided and could afford to lose the occasional shear link.

The potential for cutting larger openings was also considered. The most likely reason anticipated was the formation of stair or escalator openings, in the order of 5 m by 2·5 m. The solution adopted was to agree zones where large openings could be cut in future and to provide additional tendons in these areas, together with conventional reinforcement to deal with local stresses. This level of flexibility is not a free solution: it involves additional tendons and conventional reinforcement; discussions were therefore held with the client to agree where on the slab this would be needed.

It should be noted that the tendons used were bonded. This means that cut tendons remain effective at a distance away from the cut, and a measure of robustness is provided against accidental cutting of tendons. Despite a careful process for cutting of holes, tendons were cut accidentally a couple of times during the fit-out stage, but the capacity of the slab to take the

design loads was shown to be unimpaired.

Finally, consideration was given to the additional measures which could be taken at a later date to deal with change. Two techniques were noted, while acknowledging that they would need further design work to justify the particular circumstances. The first technique considered was the use of epoxy anchors. This would involve breaking out locally round a duct, removing the metal sleeving to expose the tendons and surrounding the tendons with an epoxy mortar. The tendon could then be cut adjacent to the epoxy mortar, which would act as an anchorage and ensure the effectiveness of a tendon near to the edge of a newly formed opening. The second technique considered was to bond carbon fibre strips to the soffit or top surface of the slab, to act as conventional reinforcement.

Optimisation for construction

BAA's approach to construction included looking for means to maximise off-site work. Two measures in particular were adopted on T5B.

The edges of the slab were detailed as precast, which had three advantages. The first was that no edge shuttering was needed—the precast units were just lowered into position. The second was that the live anchorages for the tendons could be positioned accurately under factory conditions within the edge units. The final advantage was that sockets in the top surface of the units could be pre-formed to enable edge protection to be fixed easily. One important consideration when contemplating use of such units is the availability of suitable lifting equivalent. In this case, rail-mounted tower cranes with a capacity in the order of 10 t at the appropriate radius were being provided in any case for the roof trees.

The columns were also detailed as pre-cast,

Fig. 16. Completed T5B roof steelwork on second floor slab

Fig. 17. Roof tree and foot assembly in the finished T5B departure gate

and again the lifting capacity was available to facilitate this. The column-to-lower-slab connection was detailed using part-threaded, part-grouted couplers. This meant that the column could be lowered straight into position and the bars projecting from the bottom could be dropped straight into the coupler half-full of grout.

Movement joints and creep

Conventionally, RC superstructures have movement joints at about 60 m centres. Post-tensioned structures are expected to move more because of immediate shortening and long-term creep.

In this case, movement joints were provided at 72 m centres typically, with the end-sections being 117 m long. This involved careful analysis of the relative movements, particularly between the RC ground floor and the post-tensioned first floor.

The effects were ameliorated because of the large floor-to-floor height, which helped to reduce the amount of moment attracted by the column; however partitions at ground floor level still needed careful detailing to avoid any racking effect.

Roof-build sequence

As with the roof of T5A, the assembly sequence of the T5B roof was a key driver for its structural design. The challenges of dimensional tolerances were similar to those on the T5A roof but, because the scale of the space on the T5B departures level is more intimate, the team decided to avoid potential difficulties in fit-up by welding the key connections on site.

Welding in situ can pose some serious safety risks to those who carry it out. The welding of the main nodes thus took place in a controlled environment provided by a purpose-built jig, which also functioned as an assembly shop. The programme of erection was planned around production of tubular steel roof trees in the jigs and their transport from the assembly area to the workface (Fig. 15). They were then craned onto their cast steel 'foot' nodes bolted to the slabs, after which the rafters, purlins and braces were added (Fig. 16).

Figure 17 shows one of the roof trees within the completed T5B seating area.

Conclusion

The fact that the two terminal buildings were constructed on time and to budget, despite the considerable challenges that the team faced, is a testament to the professionalism of all who were involved, and to the way that the use of a true partnering contract can yield significant benefits.

References
1. DAWSON T., LINGHAM K., YENN R., BEVERIDGE J., MOORE R. and PRENTICE M. Heathrow Terminal 5: building substructures and pavements. *Proceedings of the Institution of Civil Engineers, Civil Engineering*, 2008, **161**, Special Issue—Heathrow Airport Terminal 5, May, 38–44.
2. MCKECHNIE S., HULME P., THIND G. and MITCHELL D. Design and construction of Terminal 5 roof. *The Structural Engineer*, 2004, **82**, No. 18, 25–31.
3. FRANKLAND W., KITCHENER J. N., WHITTEN T. and HULME P. The delivery of the roof of Terminal 5. *Proceedings of the Institution of Civil Engineers, Management, Procurement and Law*, 2007, **160**, No. 3, 101–115.
4. FRANKLAND W. and HULME P. T5A roof: the abutment first run study. *Proceedings of the Institution of Civil Engineers, Structures and Buildings*, 2008, **161**, No. 2, 57–64.
5. MCKECHNIE S. Terminal 5, London Heathrow: The main terminal building envelope. *Arup Journal*, 2006, **2**, 36–43.

What do you think?

If you would like to comment on this paper, please email up to 200 words to the editor at journals@ice.org.uk.

If you would like to write a paper of 2000 to 3500 words about your own experience in this or any related area of civil engineering, the editor will be happy to provide any help or advice you need.

Proceedings of ICE

Civil Engineering 161 May 2008
Pages 54–59 Paper 700046

doi: 10.1680/cien.2007.161.5.54

Keywords
rail & bus design; railway systems;
transport planning

Ian Fugeman
BEng, CEng, FICE

was formerly BAA head of T5 rail
and tunnels

Heathrow Terminal 5: rail transportation systems

Critical to opening the new £4·3 billion Terminal 5 at London's Heathrow airport was the availability of a public transport service on extensions to both the Heathrow Express and London Underground's Piccadilly line, not least as this was a key requirement of the planning permission. A sub-surface tracked transit system also needed to be operational to provide a passenger link between the two main terminal buildings. This paper describes the challenge of managing the design, delivery and integration of new railway systems into existing, occasionally old, public systems with continually evolving management structures and without impacting on existing rail services. This was greatly assisted by the application of a system engineering approach.

The one common feature of the three rail transportation systems delivered as part of Heathrow airport's £4·3 billion Terminal 5 (T5) programme extensions to Heathrow Express and London Underground's Piccadilly line and a tracked transit system (Fig. 1) is that they were all subject to the same statutory regulations. When the project delivery teams were being

Fig. 1. Aerial view of T5 nearing completion, showing the locations of the Piccadilly line and Heathrow Express extensions and the tracked transit system

Institution of Civil Engineers

formed in the late 1990s, the primary regulations were

- Railways (Safety Case) Regulations 1994
- Railways and Other Transport System (Approvals of Works, Plant and Equipment) Regulations 1994 (ROTS)
- Fire Precautions (Sub-Surface Railway Station) Regulations 1989.

Normally the objective would simply be to achieve compliance with the regulations and codes to the satisfaction of HM Railway Inspectorate and the fire authority (initially the London Fire and Civil Defence Authority and subsequently the London and Emergency Planning Authority). The project team was faced with a greater challenge, however, which was to achieve this to the satisfaction of the various operators. Furthermore the operators, primarily owing to their differing approach to managing risk, each interpreted the regulations in a different manner.

Owing to the prolonged nature of the T5 programme, with the Transport and Works Act application being submitted in 1994 and the ultimate opening for passengers occurring in 2008, the environment in which the three transportation systems were to be delivered was continually evolving and changing. Airport owner BAA was fortunate in maintaining one office to manage the development and delivery of the three railway systems, which achieved a consistent approach by continuity.

Major evolutions and changes included the following.

- *London Underground*: the closure of London Underground Ltd under the London Transport Authority; the creation of the three shadow InfraCos in 2000; the transfer of London Underground to Transport for London (TfL); and the award in December 2000 of the public–private partnership contract for the maintenance and upgrade of the Jubilee, Northern and Piccadilly lines to Tubelines Ltd.
- *National Rail*: under the 1993 Railways Act, the old British Rail was split up and sold off. By November 1997, British Rail had been divested of all its operating railway functions. From April 1994 to October 2001, Railtrack plc was in control of the national rail infrastructure until it went into receivership and was sold to Network Rail. The Strategic Rail Authority was created in 1999 and wound up in December 2006.
- *ROTS*: process changed to Railways and Other Guided Transport Systems (Safety) Regulations 2006 under the Office of the Rail Regulator.

Another unusual feature is that the Heathrow Express service and infrastructure is owned and operated by BAA through its subsidiary company Heathrow Express Operating Company. The role of infrastructure controller is sub-contracted to Network Rail.

In addition, the contractual arrangements with the railway organisations were very different to the principles of the T5 agreement, through which BAA contracted with some 80 first-tier suppliers.[1] Typically the contractual policy of the rail industry associated with London Underground and the national rail network are based on transferring risk to the supplier at a price. On occasion this can lead to an adversarial relationship and lack of trust between operators and suppliers.

BAA's policy for the delivery of the T5 programme was the complete opposite. Thus the team accountable for the delivery of the T5 rail transportation systems was faced with bringing together operators and suppliers, often contracted in a different manner to the T5 agreement, but with the objective of working together in integrated teams in a collaborative way. On balance this was realised, with key target dates being achieved and possibly at a better value than would normally be the case.

T5 rail interchange

The rail interchange has a unique position and function within the T5 campus. During the planning inquiry, great emphasis was placed on the provision of an inter-modal facility between bus, coach, taxi, car, rail and air—a public transport interchange. This exists as a vertical penetration within the 30 m wide interchange plaza between the 4000-space multi-story car park and the west façade of the main terminal building T5A (Fig. 2).

Banks of five-person lifts serve each platform concourse and transport passengers directly from the rail platform up 32·5 m to the departures level in T5A. The lifts return to the apron level and then to the rail platforms together with escalator systems. At apron level, the interchange is covered by a 3600 m² ethylene tetrafluoroethylene roof, which permits natural light to permeate down to the rail platforms 14 m below the apron. This not only enhances passenger experience but also permits intuitive way-finding.

The station lies within a 280 m by 90 m wide box beneath T5A[3] and houses two 120 m long platforms for the Piccadilly line, two 212 m long platforms for Heathrow Express and safeguarding for a further two (see Fig. 3).

Heathrow Express extension

In its fully developed form, the Heathrow Express extension will have four platforms at T5 station, accessed via two crossovers that will permit parallel moves and the necessary turnouts. The initial construction included one crossover, permitting parallel moves, leading to two platforms (3 and 4). At initial opening in March 2008, the area of the two future platforms (1 and 2) and the tunnels leading to them were not fitted out and were segregated from

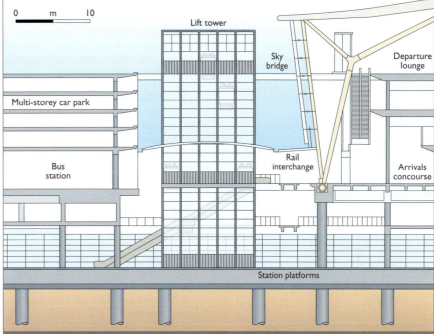

Fig. 2. Cross-section of the T5 public transport interchange, which allows natural light down on to the rail platforms: rail passengers are carried to the top departures level via lifts

(Figure labels: Lift tower; Sky bridge; Departure lounge; Multi-storey car park; Rail interchange; Arrivals concourse; Bus station; Station platforms; scale 0 m 10)

the operational railway (Fig. 3). In addition, tunnels leading to the west were constructed to facilitate and safeguard a western extension: this was a condition of the planning consent.

The main rail parameters are

- International Union of Railways (UIC) 54 running rails inclined at 1:20 except in switches and crossings
- track dynamic resistance of around 30 MN/m per rail set
- alignment suitable for 130 km/h running between Heathrow Central station and main crossovers
- alignment suitable for 70 km/h running between main crossovers and station area
- alignments suitable for 50 km/h running into T5 station
- alignment suitable for 35 km/h in platform areas
- track work to support a train service of up to 16 trains/h (equivalent tonnage 22 MGT/a)

The track form was considerably modified from that used on the original Heathrow Express (Fig. 4). The derailment containment upstand was moved to outside the running rails, and the track slab lowered relative to the underside of the rail. Base plates were used to provide vertical adjustability and greater resilience.

The Heathrow Express extension rail alignment was developed to accommodate two possible future western extension rail schemes. Of the two schemes, Airtrack is the most developed and is planned to continue in tunnel to the west, curve to the south west and surface south of junction 14 on the M25 motorway. The T5 project re-assessed an earlier British Rail concept scheme, undertook a multi-disciplinary feasibility study and addressed emergency scenarios, ventilation and environmental considerations. This was necessary to ensure that if and when Airtrack proceeds, its execution will have no

impact on the Heathrow Express or airport operations. Prior to starting construction of T5, the rail alignment was further modified to accommodate safeguarding a rail extension to the north west, either in association with or independently of the well-defined Airtrack scheme.

The rail environment, including the T5 station, lies within 763 m of cut-and-cover concrete box structures. The remainder of the route is within a 5·675 m diameter bored tunnel of a length of 1·45 km down-line and 1·87 km upline.[2] An emergency access and egress shaft was built next to the future second satellite terminal T5C, and is combined with a similar facility for the adjacent Piccadilly line extension.

Traction power is provided via overhead line equipment fundamentally the same as the original Heathrow Express but with additional clearances achieved by reducing the depth of the trackbed. The design and installation of the overlap and interface with the existing Heathrow Express was achieved with no disruption to the existing service to the central terminal area (Terminals 1, 2 and 3) and Terminal 4.

The signalling and signal control systems were originally conceived as more of the same: an extension of the existing system operated on BAA's Heathrow Express system serving central terminal area and Terminal 4 stations, which was brought into operation in May 1998.

It was always known that the existing configuration at the Slough integrated electronic control centre utilised solid-state interlocking, which was at full capacity. The logical solution was to create additional capacity at Slough. A solution could not, however, be found which was acceptable to the asset owner (Railtrack and subsequently Network Rail) and consequently options were developed by which the signalling control and interlocking would note be located on Network Rail's property, which logically meant it would no longer act as infrastructure controller on behalf of the owner and operator of Heathrow Express.

By December 2004 BAA was minded to award

a contract to Siemens for a computer-based interlocking signalling system to be located at Heathrow, with the operating company taking back the role of infrastructure controller. Network Rail's most senior management suggested that BAA might like to reconsider if Network Rail assisted in finding a technical solution by which the additional capacity could be created within the existing Slough control centre. The offer was accepted, and during 2005 BAA contracted Network Rail via a specific implementation agreement to manage the design and delivery of the signalling system. BAA, with its first-tier suppliers engaged via the T5 agreement, sought close integration of Network Rail's suppliers engaged via traditional lump-sum contracts.

Slough control centre previously had two signaller workstations, each controlling six solid-state interlockings. The maximum number of interlockings per workstation was historically six and the maximum number of interlockings per control centre was 12. The novel solution was the addition of a new interlocking, which required the provision of a new second control centre allocated to the existing second workstation, and an alteration to the system to allow up to seven interlockings per workstation (subject to the maximum of 12 interlockings per control centre).

The Heathrow Express extension branch is signalled bi-directionally to provide operational flexibility and required 18 new signals. There are ten main signals, four repeater signals and four banner signals. All signals are searchlight light-emitting diode (LED) type. The existing automatic warning system installed on the Heathrow Express has also been extended together with automatic train protection using the existing Alstom ACEC equipment.

Piccadilly line extension

The Piccadilly line was first extended to the Heathrow central terminal area in 1977 and further extended to Terminal 4 in 1986. The

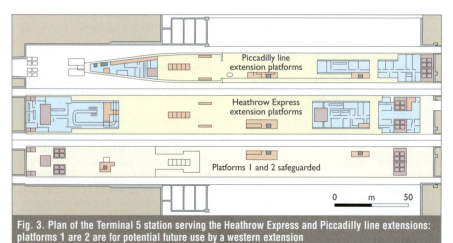

Fig. 3. Plan of the Terminal 5 station serving the Heathrow Express and Piccadilly line extensions: platforms 1 are 2 are for potential future use by a western extension

Fig. 4. Heathrow Express train at the new Terminal 5 station: track form on the extension is a considerable modification to that on the rest of the route

direct service from central London was to Hatton Cross followed by Terminal 4 and then, via a loop tunnel, to Terminals 1, 2 and 3 prior to returning to Hatton Cross (Fig. 5). The extension to Terminal 5 (Fig. 6) was connected to the existing loop immediately to the east of the location where the existing single loop tunnel bifurcates through a step plate junction into two tunnels a few hundred metres to the west of the central terminal area.[2]

Similar to the Heathrow Express extension, the Piccadilly line extension comprises two single-track bored tunnels with cut-and-cover structures on either side of the T5 station. Twin reversing sidings with a diamond crossover to the west of the station provide the required operational functionality and are also designed so as not to preclude a future extension to the west.

The Piccadilly line service is now split between the existing route around the Terminal 4 loop and the new T5 spur, which runs Hatton Cross—Terminals 1,2,3—T5—Terminals 1, 2, 3 —Hatton Cross.

The most unusual feature of the Piccadilly line extension was the procurement route. In August 2004 a contract agreement was reached between Heathrow Airport Ltd and London Underground for delivery of the Piccadilly line extension works and ongoing maintenance obligations for 30 years following its opening for passenger services. This quasi-private-finance-initiative contract had taken some 6½ years to negotiate since the signing of a memorandum of understanding, which established London Underground as client and Heathrow Airport as a contractor bearing all construction cost and risk.

In return, London Underground paid the airport a share of the incremental passenger revenue over a 30-year period. The restructuring of London Underground in 2000, however, meant that elements of the Piccadilly line extension had to be delivered by its private-finance-initiative and long-term contractors, with which the airport could have no contractual relationship. These works were procured by London Underground via a variation with its contractors. The airport remained liable for

all cost and risk arising from the performance of these contractors, even though there was no direct contract. The fact that some very complex integration works occurred successfully and to programme is remarkable, especially considering that the two major parties each had procurement strategies and attitudes to risk management which were diametrically opposite.

Traction power

For obvious reasons of compatibility with rolling stock, the traction power supply arrangement for the extension is an extension of the existing four-rail 630 V direct-current system. There was an initial assumption, based on common practice, that there should be a new 'end of track' rectifier substation at T5 to supply the extension traction load. Modelling demonstrated that the existing rectifier substations at Terminals 1, 2, 3 and Hatton Cross could, however, sustain the required traffic pattern, while still allowing for credible plant outage. Having agreed this basic arrangement, there were no significant technical challenges and standard switchgear equipment, already in use elsewhere on the London Underground network, was selected for the installation.

The modelling also showed that composite conductor rail, made from aluminium with a bonded stainless-steel top surface, had reduced resistance compared to conventional steel rail and should be used to ensure adequate voltage at trains in all locations. The lighter weight of composite rail and its bolted (no welding) construction also brought significant installation advantages.

The Piccadilly line extension eastbound and westbound tracks are fed as separate traction sections from the main direct-current busbars at Terminals 1, 2, 3 station, but these are normally connected together by track-coupler breakers at T5. This gives 'pseudo-double-end' feeding, with reduced volt drop under normal operation. To enhance resilience and reliability of supply further, the supply circuit breakers at Terminals 1, 2, 3 and the coupler breakers at T5 are all duplicated: one breaker from each pair can be out of service for maintenance without disruption to

traffic. The traction power installation at T5 also includes switches that can be opened to remove power from each of the two reversing sidings, allowing a faulty train to be isolated.

The new direct-current switchgear is equipped with direct-acting and electronic fault-protection equipment, to current standards. Full remote control and indication of all the new equipment is provided at London Underground's power control room, completely integrated with the existing Piccadilly line traction-supply systems.

Signalling and signalling control

The signalling systems on the Piccadilly line are typical of those on much of the London Underground network, comprising

- mechanical lever frame interlocking
- two-aspect colour light signalling
- pneumatically powered points and train stops (a simple train protection system)
- signalling control systems utilising computers of 1970s vintage.

The key challenges in signalling the T5 extension were

- modifying and interfacing to existing systems
- providing modern equivalents of legacy systems
- minimising the extent of approvals required
- providing a points solution and safety case that addressed the issues arising from the Potters Bar accident
- integrating the signalling with the track design, particularly the BB-54m shallow depth switch
- integrating the new signalling control system into the existing control centre at Earls Court.

To this end, the following systems were implemented.

- Westinghouse-style 63 electric point machines were provided to drive the new

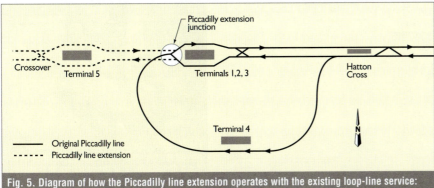

Fig. 5. Diagram of how the Piccadilly line extension operates with the existing loop-line service: trains from Hatton Cross will alternate between Terminal 4 and Terminal 5

Piccadilly extension junction

Crossover   Terminal 5   Terminals 1,2, 3   Hatton Cross

Terminal 4

N

—— Original Piccadilly line
- - - - Piccadilly line extension

Fig. 6. Piccadilly line train at the new Terminal 5 station

point ends at the new junction and the T5 crossover. Additional detection at back end of the switch was provided, which was novel to London Underground, giving improved safety.

■ A new relay interlocking was implemented at T5, based on designs provided for the Central line in the 1980s. This interlocking-controlled external signalling equipment is similar to that on the remainder of the Piccadilly line.

■ The lever-frame interlocking at Terminals 1, 2, 3 was modified to enable control of the two new point ends at the Piccadilly line extension junction (300 m west of Terminals 1, 2, 3) to enable access and egress from the existing railway to the extension.

■ A new signalling control system based on a programmable logic controller was provided at Earls Court control centre and at the T5, Terminals 1, 2, 3 and Hatton Cross stations to control the new systems and implement the new services. The new system was required at Hatton Cross to enable trains to be timetabled and controlled to both routes (Terminals 1, 2, 3 via Terminal 4 and T5 via Terminals 1, 2, 3).

All of the systems were designed in-house by Tube Lines and required a larger element of modification to existing systems relative to the new elements being introduced.

Ventilation

The tunnel ventilation systems follow the principles used for the design of ventilation on Heathrow Express.[4] The Heathrow Express extension has a ventilation shaft approximately half-way between Terminals 1, 2, 3 and T5. The Piccadilly line extension has a similar shaft slightly farther west, at the planned T5C terminal site. These two shafts provide push–pull ventilation during emergency situations in the tunnels between T5 and the two existing railways.

In the T5 station, the ventilation system consists of draught relief and forced ventilation.[5] The fans at the station are shared between Piccadilly line and Heathrow Express extensions, based on the assumption that they will never be needed to deal with fires on both railways at the same time. The Heathrow Express platforms are provided with under-platform exhaust switched by thermostats on the platforms, while London Underground's reversing siding is kept cool by continuous forced ventilation. The terminal building is air-conditioned, so it is separated from the station environment by dividing walls and lobbied, sliding doors.

The nature of the T5 site meant that all ventilation outlets at the station had to be placed at the west end of the station, within the multistorey car park. Ventilation inlets at the east end of the platforms are connected to the outlets by air ducts running beneath the platforms. These connect to a manifold of air ducts above the station, which are built around the massive beams that support the car park. These beams are 5 m deep, spaced 9 m apart and span 80 m, so there is ample space between them for the air ducts. The manifold connects to draught relief within the car park and to four ventilation fans, each within one of the four cores of the car park's spiral ramps (see Fig. 1).

Glass walls separate the rail platform from the station concourse with openings for passenger access. In the unlikely event of a train fire at the station, the ventilation will extract air from both ends of the incident platform. The make-up air is drawn through the openings in the glass wall at a fast enough rate to contain smoke in the incident platform. The intention is that any passengers walking along a smoke-filled platform away from a fire will quickly encounter a stream of clean air coming out onto the platform through the access route, and will follow it into the station concourse. The glass walls also enable visual connectivity across the two rail systems.

Table 1 gives brief details of the ventilation fans used in Heathrow Express and Piccadilly line extensions.

Tracked transit system

The T5 tracked transit system forms part of the passenger horizontal and vertical circulation for T5, linking the main terminal T5A to the satellite terminals. Vertical circulation, in the form of lifts and escalators, transports passengers from the arrivals and departures levels of the buildings to the tracked transit system subsurface stations (or passenger walkways below for recovery passengers or in contingency). The stations are located beneath each concourse building in an east-to-west orientation (Fig. 7), and the horizontal circulation is provided by dedicated driverless trains, or 'automated people movers' (Fig. 8).

The system was designed to be built in phases. Phase 1 was required to be operational at the time of opening in March 2008 and enable passengers to travel from T5A to T5B, an approximate travel distance of 280 m. Phase 1 also includes train functionality to the future T5C and a maintenance facility. This mitigates disruption to the operational service post handover of phase 1.

The subsequent phase 2 includes creation of T5C and requires the enablement and fit-out of the associated tracked transit station, notionally in 2010. Further design consideration has been

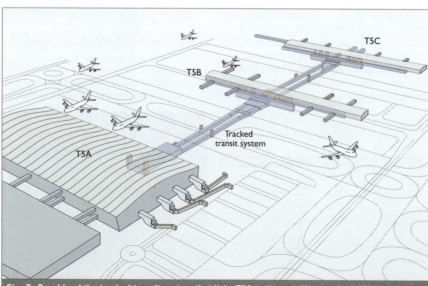

Fig. 7. Graphic of the tracked transit system that links T5A to the satellite terminals T5B and T5C via lifts, escalators and sub-surface driverless trains

Fig. 8. Bombardier Transportation Innovia driverless trains in the tracked transit system are a development of those in use at Gatwick and Stansted airports: they are initially running every 90 s over a 20 h period

given with regards alignment, for expansion to a third satellite, T5D, and/or to the Heathrow central terminal area. Peak flow rates on the phase 1 of the tracked-transit system can be 5800 passengers per hour in each direction. The service interval will alternate between each guideway, with a train frequency of every 90 s.

In phase 1, the trains runs in two sub-surface running tunnels (cut-and-cover construction) below the aircraft apron area. These tunnels are equipped with a dedicated ventilation system to provide general air movement, draught relief for the piston effect of trains moving through tunnels and a safe environment for personnel access and egress of tunnels during incident. Directly below the running tunnels are two parallel walkways allowing an alternative route between T5A, T5B and T5C. For security reasons, the walkways are each dedicated for flight arrivals and departures and have been sized to accommodate electrically powered vehicles for mobility impaired passengers, pedestrians and the future provision of passenger conveyors.

The tracked transit system consists of two parallel independent guideways designed to operate as a dual shuttle system in phase 1, with the capability of operating as a shuttle or pinched loop in future phases. During phase 1 the system will consist of two trains, one on each guideway travelling independently back and forth between T5A and T5B over a 20 h period in a dual shuttle configuration on the two guideways

To allow for future expansion to a pinched loop, switches have been installed in phase 1 between the guideways. In phase 2, pinched-loop mode, each train consisting of one or more vehicles will travel up one guideway, serving the stations on that guideway. The crossovers—a combination of four pivot switches and one turn-table switch between the guideways at turn-back stations—allow the trains to transfer to the other guideway and serve the platforms on that guideway. In this way up to four trains will be able to follow one another around the track in a continuous loop operation.

Vehicles, track and power supply

The tracked transit system trains are made up of a configuration of one to four electrically powered rubber-tyred vehicles that can be coupled together to form differing train lengths. Each vehicle is guided by wheels running against a central steel guide beam running along a flat concrete guideway.

Propulsion power to the vehicles is collected from two 350 V direct current rails, plus earth rail, mounted on the guide beam. It is controlled onboard to drive single 750 V alternating current traction motors. The vehicles have a maximum line speed of 50 km/h.

The vehicles are a new Bombardier Transportation Innovia product, which is the next evolution of the proved C/CX-100 technology in service at Gatwick and Stansted airports. Unlike the C/CX100, the vehicles deliver a lower profile design since the guidance structure is level with the running surface. This results in smaller running-tunnel height and thus a shallower structure. The vehicles use the latest automated train control, protection and supervision technology for vehicle location and headway control. Known as Cityflow 650, this utilises a radio-communication-based moving-block train-control system in lieu of fixed-block control, providing less wayside equipment, more flexibility and easier expansion.

Each vehicle is equipped with four sets of automatically operated bi-parting doors, two on each side of the vehicle. During a station stop, the vehicle doors align with similar door sets on the platform edge which are interlocked. The vehicles have been designed to meet the latest standards for fire safety performance and the interiors are fully accessible to meet UK legislation. The station platforms have step-free access throughout.

Stations and maintenance facility

The tracked transit system stations and passenger walkways are served by vertical circulation, which carries passengers down to and up from the stations and passenger walkways. Passengers elect to use to use either lifts or escalators.

For flight arrivals and departures security segregation, each station consists of three platforms, with the central platforms having two platform edges. Each platform edge is of a length to accommodate eight sets of bi-parting doors (two sets per vehicle berth position).

Each station has two arriving and two departing platform edges. Departing passengers at T5A board a train from one of the two platform edges between the guideways. The train travels to T5B, where passengers alight onto platforms on the outside of the guideways. The adjacent boarding and alighting of trains means that pas-sengers move across a vehicle without conflicting flows commonly experienced on railways. It also achieves the security requirement to segregate arriving and departing passengers.

A 3500 m² maintenance facility has been built to the east of the T5C site, which has two light-maintenance train bays with access pits and a heavy-maintenance bay for jacking vehicles. The facility can accommodate up to 12 vehicles at any one time and has support accommodation to allow for off-line 24/7 maintenance.

Conclusion

All three new rail transportation systems at Heathrow T5 encompassed a wide variety of technical systems, each with their own challenges. In addition to the management of interfaces was the challenge of integrating new systems with existing without impact on ongoing operations. System integration processes and the principles of system assurance were applied from the start to ensure the successful delivery of an assured product to the satisfaction of the various asset owners, operators and statutory bodies.

References

1. WOLSTENHOLME A and FUGEMAN I. Heathrow Terminal 5: delivery strategy. *Proceedings of the Institution of Civil Engineers, Civil Engineering*, 2008, 161, Special issue—Heathrow Airport Terminal 5, May, 10–15.
2. WILLIAMS I. Heathrow Terminal 5: tunnelled underground infrastructure. *Proceedings of the Institution of Civil Engineers, Civil Engineering*, 2008, 161, Special issue—Heathrow Airport Terminal 5, May, 30–37.
3. DAWSON T., YENN R., BEVERIDGE J., MOORE R. and PRENTICE M. Heathrow Terminal 5: building structures and pavements. *Proceedings of the Institution of Civil Engineers, Civil Engineering*, 2008, 161, Special issue—Heathrow Airport Terminal 5, May, 38–44.
4. GRAY W. G., SWEETLAND I. and VAN VEMDEN F. Tunnel ventilation for the Heathrow Express link. *Proceedings of the 9th International Symposium on the Aerodynamics and Ventilation of Vehicle Tunnels*. BHR Group, 1997.
5. GRAY W. G., BENNETT E. C. and GETTINGS L. Design of the tunnel ventilation system at the Terminal 5 station. *Proceedings of the 11th International Symposium on the Aerodynamics and Ventilation of Vehicle Tunnels*. BHR Group, 2003.

What do you think?

If you would like to comment on this paper, please email up to 200 words to the editor at journals@ice.org.uk.

If you would like to write a paper of 2000 to 3500 words about your own experience in this or any related area of civil engineering, the editor will be happy to provide any help or advice you need.

Table 1. Ventilation fan details

Location	No. of fans	Duty flow	Total pressure rise
Heathrow Express extension T5D shaft	2	140 m³/s	1125 Pa
Piccadilly line extension T5C shaft	2	110 m³/s	610 Pa
T5 station	4	280 m³/s	1300 Pa
Piccadilly line extension reversing siding	1	25 m³/s	135 Pa

Proceedings of ICE
Civil Engineering 161 May 2008
Pages 60–64 Paper 700047

doi: 10.1680/cien.2007.161.5.60

Keywords
airports; district heating;
power stations (fossil fuel)

George Adams

Spie Matthew Hall

Heathrow Terminal 5: energy centre

Heating and cooling the vast new terminal buildings at London Heathrow airport's £4·3 billion Terminal 5 development is undertaken by a dedicated energy centre, which provides continuous supplies of hot and chilled water for heating and air-conditioning respectively. The hot water is provided by a combination of a local combined-heat-and-power source and natural gas boilers, which can also run on fuel oil, and the chillers are powered by high-voltage electricity. This paper describes the design and construction of the highly efficient centre, which also made extensive use of off-site testing and manufacturing.

The £4·3 billion Terminal 5 (T5) development at London's Heathrow airport consists of a series of large buildings and infrastructure with their own utility service feeds. However, all heating water, cooling water and domestic water is supplied by a central energy centre (Fig. 1).

Centralised distribution of these services to the main terminal building and its satellites and ancillary areas was considered to be the most effective solution based on construction, commissioning, operations and energy efficiency considerations. It also enables future flexibility in terms of spatial allowances and utility services capacities.

The energy centre project involved large-scale off-site modularisation of mechanical and electrical services and supporting structure, and included fast-track on-site assembly methods using strand-jacking technology. It also provided an early opportunity to test the T5 infor-mation technology network and control systems integration.

The principal contractor for delivering the project was Amec, which worked in conjunction with other contractors, principal designer PB Power and client BAA's technical leadership team.

Definition and project requirements

When engineering design for the energy centre started in 1998, assessments were made on the potential use of renewable energy and combined heat and power (CHP). It was concluded T5 should not have a dedicated CHP system but that a single CHP plant serving the T5 and central terminal area (Terminals 1, 2 and 3) would be more appropriate – though in the event this did not proceed. However, during the commissioning of the energy centre, it proved possible to integrate a CHP connection from the nearby

Fig. 1. The T5 energy centre (left) provides all heating and chilling water for the new terminal buildings and ancillary areas

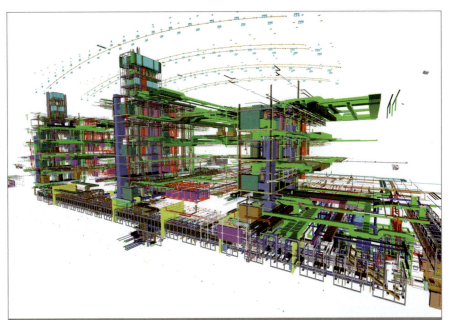

Fig. 2. Design of the energy centre required close cooperation with the services design of the terminal buildings to ensure operational heating and cooling demands were met

Thames Valley Power plant that also supplied the airport cargo area.

The scope of the energy centre project comprised the following

- centralised hot water
- centralised chilled-water plant
- heating, cooling and associated system distribution networks
- gas supplies
- high-voltage electrical supplies
- borehole water treatment plant
- heat-rejection facilities
- controls and information technology networks
- CHP supplies
- support systems such as lighting and fire alarms.

The building design was developed to the T5 site design guidelines and the solution was rigorously reviewed by the T5 design expert group. The design was also influenced by the strategic restriction on height and the then-new Building Regulations. In addition to the main plant rooms, the building includes a control room, stores area, workshop, offices and support accommodation for operator Heathrow Airport Ltd.

The overall design encompassed the heating and cooling demands for the various terminal buildings, which required close cooperation with other design teams across the T5 project. The design process established the optimum operating temperatures, pressures and controls philosophy to reduce life-cycle costs. A full energy metering system has thus been incorporated to monitor consumptions of all service provisions.

Delivery of the engineering systems was largely based on build off-site technology. Large integrated sections of the works – incorporating mechanical and electrical systems and equipment and, where appropriate, structural elements of the building framework – were manufactured and delivered to site as complete modules.

In parallel, important elements of off-site testing were integrated into the process to assist in verification of plant and equipment performance. This also enabled the incorporation of manufacturers' input to both design and fabrication processes, ensuring good technical coordination, physical fit and parallel manufacturing sequences.

Basis and standards of design

The design of the energy centre was carried out to the client's engineering guidelines, EN standards and other appropriate industry guidance such as Building Services Research and Information Association reports, Chartered Institution of Building Services Engineers guidelines and energy best-practice publications.

Systems design was based on the fundamental criteria of

- energy provision is in the form of hot water for heating and chilled water for air-conditioning
- provision from a single energy centre.

The key requirements for heating and cooling demands were for business continuity of T5 operations, which were established by the design teams for each separate building (Fig. 2). The energy centre designers were provided with temperature and pressure criteria based upon strategic considerations for the complete development.

The design took account of the hydraulic design for the pipework to each building and made extensive use of computational fluid dynamic software and three-dimensional modelling systems to prepare calculations and assessments. The complete process of design verification and sign off was conducted via project procedures including quality plans, risk assessments and technical audits. This ensured the complete design-to-commissioning process was documented, transparent and well structured.

The primary energy supply to the centre was selected as natural gas for heating and high-voltage electricity for power. It was decided that aviation fuel should be incorporated as a standby alternative for the boilers, though this was later changed to fuel oil. In basic terms, the energy centre complex converts simple primary energy sources into thermodynamically useable supplies for the T5 buildings.

Detailed design development

The detailed design stage of the energy centre incorporated wider considerations of manufacturing, construction and commissioning. Key members from the design definition stage were incorporated into this stage, enabling buildability advice to be taken into account and ensuring learning from the design process was effectively taken forward.

Some of the key deliverables from the detailed design development stage were

- detailed activity programmes, inclusive of build off-site technology
- milestone definitions, including those from the design definition stage
- change-management assessments
- detailed risk-management assessments
- buildability reports
- cost and time evaluations
- testing and commissioning plans
- procurement documentation
- three-dimensional drawings and fabrication details
- handover plan
- maintenance considerations and associated technical reviews, involving appropriate client representatives
- development of the construction health and safety plan.

The philosophy through each stage of the programme was to eliminate health and safety risks during construction, commissioning and operations. Consideration of the project's eventual demolition was also included.

Technical content

The energy centre is located to south west corner of the T5 site (Fig. 3). The heat rejection equipment located within the roof (Fig. 4) and the main plant is located at the ground floor level (Fig. 5), with distribution occurring at mezzanine level.

The interconnection is via a series of below-ground tunnels housing the following systems

- chilled-water pipes
- heating-water pipes
- potable-water pipes
- grey-water pipes
- high-voltage power cables (Fig. 6).

Table 1 provides full details of the plant and equipment incorporated in the energy centre.

To support the main plant, the hydronic pumping arrangements are as follows.

- *Heating.* Primary heating pumps and circuit serves the boilers with secondary transfer pumps to the main terminal T5A, satellite terminal T5B and future satellite T5C. There are secondary exchange plant rooms within the terminal buildings housing interface-plate heat exchangers, which step down the primary supply temperature from 95°C to 72°C, with secondary pumps within each building for general distribution.
- *Chilled water.* There are primary pumps associated with the chillers, coupled up by a common primary pipework circuit. Secondary pumps in the energy centre circulate the chilled water directly to each point of use in each terminal building.

Delivery philosophy

Following the design definition and design development stages, the delivery of the energy centre project was conducted through the following key stages

- fabrication design
- manufacturing
- assembly
- commissioning
- integration
- handover.

The philosophy was to give due consideration to each of the above stages and achieve good integration between designers, manufacturers, engineers, and project management functions co-located on the T5 site. All parties were working to the unique and open style of the T5 agreement, which sets out how all parties should work together to common objectives and deliverables.[1]

Figure 7 shows one of the five prefabricated mezzanine floor modules being stand-jacked into position inclusive of integrated mechanical systems. The benefits of the build off-site methodology are as follows.

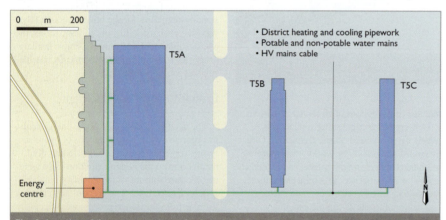

Fig. 3. Location of the energy centre and sub-surface pipework tunnels

Fig. 5. Two of the four 6·6 MW, twin compressor chiller units on the ground floor of the energy centre

Fig. 4. Aerial view of energy centre showing the four pairs of chiller cooling towers in the roof

Fig. 6. Sub-surface tunnels contain district heating and cooling pipes, potable and grey-water mains and high-voltage power cables

- *Safety.* Installing mechanical pipework up to 600 mm diameter by traditional means at high level would have significantly increased construction health and safety risks.
- *Time.* Jacking of the 6 m by 30 m sections as shown in Fig. 7 is carried out in a matter of days as opposed to many weeks of traditional construction methods. The overall benefit was calculated to be around four months for the complete project.
- *Interfaces.* The major modules incorporated steel beams that not only supported the modules but also became part of the building structural frame, reducing the amount of steelwork needed.
- *Cost.* Overall cost benefits were achieved, by reducing the length of the critical path in the programme and allowing earlier delivery of key plant items to site due to a faster availability of associated space under the mezzanine floor.

The use of three-dimensional modelling was an important feature within the delivery process. Not only was this used to develop fabrication and assembly details, but the model visualisations facilitated better risk assessments, improved understanding of the assembly

sequences and a better method of communicating to all parties involved.

Energy demand

The demands from the terminal buildings are varied and quite dramatic in their seasonal pattern. Fig. 8 contains the main cooling demand curve, which shows how the peak changes dramatically from low to high. Whereas two or three compressors are needed for the majority of the cooling season, this rapidly increases to the full eight compressors at peak condition for a relatively shorter period of time. The controls and pipework strategy accommodates these variations.

To minimise the energy consumption by the energy centre, the following technology has been incorporated into the design

- efficient plant selections
- heat recovery from the lead boiler
- incorporation of heat supply from CHP plant
- efficient pump selections with energy-saving controls incorporated
- high-efficiency plate heat exchangers
- heat exchanger coupled to one dedicated tower/refrigeration plant configuration to

provide an amount of free cooling facility (i.e. using this tower to cool the return chilled-water line from the terminal buildings when appropriate to utilise the facility)
- variable-speed drives on all associated pumps.

Commissioning and handover

The commissioning process was put in place during the design development phases. The process was based upon a number of key stages delivered through 'authority-to-proceed gateways'. These gateways enable the integration of

Fig. 7. Prefabricated 6 m by 30 m mezzanine floor module being stand-jacked into position – the process reduced high-level working and saved four months' installation time

Plant / equipment / systems	Quantity	Description
Refrigeration plant/chilled store	4/1	Chiller plants each produce up to 6·6 MW of cooling capacity, making a total capacity 26·4 MW. They are of absorption type using ammonia (R717) as refrigeration medium. Each has two screw compressors providing chilled water at 5·5°C, which after main losses is received in each building 6·6°C. Coupled with common pipe network, the 4300 m³ chilled water store has a maximum charge and discharge rate of 4 MW. Each machine is contained in its own fire control compartment.
Cooling towers	8	Heat rejection from chillers is via eight open-type cooling towers located at roof level and positioned to minimise any impact on water vapour plumage on airport operations. Open towers are thermodynamically efficient and take up less space than other types of heat rejection equipment, minimising the footprint of the building. Parallel heat exchangers used to enable cooling tower condenser water pipework to be coupled to each chiller.
Boilers	3	Boilers are located in a separate plant room section. Lead boiler is 9 MW design load and other two lag boilers are 18 MW. Lead boiler takes advantage of flue-gas heat recovery from waste heat. However, CHP was introduced later (see below).
Water treatment plant/system	4	Heating, cooling and condenser pipework all have 'water treatmatic' systems which automatically regulate water condition to avoid corrosion and bacterial growth. Cooling towers have make up of water losses supplied from borehole supply via dedicated treatment and filtration plant in the energy centre.
Supporting ventilation systems	Various	Boiler vent system is for combustion and heat relief and is facilitated by natural-ventilation louvres in main plant-room doors. Chiller ammonia-extract system is a fully mechanically assisted arrangement with variable air-volume control box located in each individual chiller enclosure. It is activated by ammonia-detection sensors linked into the fire-detection system. Ventilation system is used in a normal operating mode to provide removal of heat from the rooms.
Low-voltage distribution system	1	Low-voltage power distribution system is a standard configuration complying with specific client guidelines. The four chillers are directly fed primarily by high-voltage power supply to improve operational efficiency.
Fire-detection and alarm system	1	General fire-detection system incorporates heat detection in boiler room and ammonia detection in each of the chiller enclosures.
Lighting/lighting control system	1	General lighting throughout to suit individual areas, coupled with emergency lighting back-up. A control system is incorporated to avoid general lighting operating unnecessarily.
System-integration network	1	T5 common information technology network manifests itself in the energy centre in an 'access head end' computer. Engineering systems are graphically represented so that operating conditions, alarms and trends can be reviewed.
Engineering controls systems	Various	Heating, cooling, ventilation systems all locally controlled in the energy centre by dedicated control outstations and networks, integrated into the T5 network. All control software was tested off-site in a dedicated facility to review peer-to-peer system communications and integrity.
CHP heat-injection system	1	Late in the programme, an opportunity arose to include heating supply from combined heat and power plant/system located independently on the southern side of the airport. This source can provide up to 15 MW of heat at a temperature to achieve, via a separation plate heat exchanger, the required T5 system temperature of 95°C. As a result, the control strategy was refined during commissioning to enable the CHP supply to take up the lead heat supply source, with boiler 1 taking the next sequenced heat source. This arrangement has significant benefits in operation.
Fuel-oil standby system	1	Late in the programme, the client supported a proposal to change from aviation fuel as the standby energy source to standard 35 second oil locally stored. This alleviates dependency on use of aviation fuel, over which aircraft operations take priority.

Table 1. Summary of plant and equipment in the energy centre

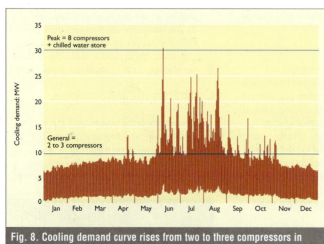

Fig. 8. Cooling demand curve rises from two to three compressors in winter to all eight plus the chilled water store in the summer

Fig. 9. The commissioning and handover process was based upon a number of key stages delivered through 'authority-to-proceed gateways', ensuring the client's operational staff could operate and maintain the systems effectively and safely

a number of related sub-systems into a common and auditable process (Fig. 9).

Each of the authority-to-proceed gateways incorporated a detailed schedule of requirements that had to be satisfied and signed off prior to the move to the next commissioning stage. The client arrived at the handover stage, having been involved in every preceding section of the delivery process.

The handover plan was not just about a formal transference of responsibility and information, the process encompassed all of the following

- record information and drawings
- maintenance documentation
- training schedules for client's operations staff
- operational training for systems
- risk residuals
- witnessing of commissioning results and site acceptance tests
- sign off inspections by the building control representative
- asset coding schedules.

All of these were coordinated within a specifically drafted handover agreement, so that a summary of all the documentation references is located in place.

The key to the handover was to develop the client's operational staffs' knowledge and information so that they could operate and maintain the systems effectively, safely and with understanding of the dynamics of the engineering solutions.

Conclusion

The energy centre lies at the heart the £4·3 billion T5 development, ensuring the effective and sustainable supply of energy services to the terminal buildings (Fig. 10).

The robustness of the design, build and commissioning processes has delivered a project that provides critical business continuity for the airport operator, airline operators and retail concessionaires.

Reference

1. WOLSTENHOLME A., FUGEMAN I. C. D. and HAMMOND B. Heathrow Terminal 5: delivery strategy. *Proceedings of the Institution of Civil Engineers, Civil Engineering*, 2008, **161**, Special issue—Heathrow Airport Terminal 5, May, 60–64.

What do you think?

If you would like to comment on this paper, please email up to 200 words to the editor at journals@ice.org.uk.

If you would like to write a paper of 2000 to 3500 words about your own experience in this or any related area of civil engineering, the editor will be happy to provide any help or advice you need.

Fig. 10. Completed energy centre ensures effective and sustainable supply of energy services to the new T5 terminal buildings